Conservation Management of Freshwater Habitats

CONSERVATION BIOLOGY SERIES

Series Editors

Dr F.B. Goldsmith
Ecology and Conservation Unit, Department of Biology, University College London, Gower Street, London WC1E 6BT, UK

Dr E. Duffey OBE
Cergne House, Church Street, Wadenhoe, Peterborough PE8 5ST, UK

The aim of this Series is to provide major summaries of important topics in conservation. The books have the following features:

- original material
- readable and attractive format
- authoritative, comprehensive, thorough and well-referenced
- based on ecological science
- designed for specialists, students and naturalists.

In the last twenty years **conservation** has been recognized as one of the most important of all human goals and activities. Since the United Nations Conference on Environment and Development in Rio in June 1992, **biodiversity** has been recognized as a major topic within nature conservation, and each participating country is to prepare its biodiversity strategy. Those scientists preparing these strategies recognize **monitoring** as an essential part of any such strategy. Chapman & Hall has been prominent in publishing key works on monitoring and biodiversity, and aims with this new Series to cover subjects such as conservation management, conservation issues, evaluation of wildlife and biodiversity.

The Series contains texts that are scientific and authoritative and present the reader with precise, reliable and succinct information. Each volume is scientifically based, fully referenced and attractively illustrated. They are readable and appealing to both advanced students and active members of conservation organizations.

Further books for the Series are currently being commissioned and those who wish to contribute to the Series, or would like to know more about it, are invited to contact one of the Editors or Chapman & Hall.

Already Published

1. **Monitoring Butterflies for Ecology and Conservation**
   Edited by E. Pollard and T.J. Yates (Hb 1993 o/p, Pb 1995)
2. **Insect Conservation Biology**
   M.J. Samways (Hb 1994 o/p, Pb 1994)
3. **Monitoring for Conservation and Ecology**
   Edited by F.B. Goldsmith (Hb/Pb 1991, Pb reprinted four times)
4. **Evaluation and Assessment for Conservation: Ecological guidelines for determining priorities for nature conservation**
   I.F. Spellerberg (Hb 1992 o/p, Pb 1994 reprinted three times)
5. **Marine Protected Areas: Principles and techniques for management**
   Edited by S. Gubbay (Hb 1995)
6. **Conservation of Faunal Diversity in Forested Landscapes**
   Edited by R.M. DeGraaf and R.I. Miller (Hb 1996)
7. **Ecology and Conservation of Amphibians**
   T.J. Beebee (Hb 1996)
8. **Conservation and the Use of Wildlife Resources**
   M. Bolton (Hb 1997)
9. **Conservation Management of Freshwater Habitats**
   P.S. Maitland and N.C. Morgan (Hb 1997)

Forthcoming

Primate Conservation Biology
G. Cowlishaw and R.I.M. Dunbar
Coastal Management for Nature Conservation
P. Doody
Wetland Ecology and Management
B.D. Wheeler
Valuation of Costs and Benefits of Wildlife in Africa
Edited by H.H.T. Prins and J.G. Grootenhuis

Conservation Management of Freshwater Habitats

Lakes, rivers and wetlands

P. S. Maitland
Fish Conservation Centre, Gladshot, Haddington, Scotland

and

N. C. Morgan
7 Route de l'Envers, 88290 Thiéfosse, France

Published by Chapman & Hall, 2–6 Boundary Row, London SE1 8HN, UK

Chapman & Hall, 2–6 Boundary Row, London SE1 8HN, UK

Chapman & Hall GmbH, Pappelallee 3, 69469 Weinheim, Germany

Chapman & Hall USA, 115 Fifth Avenue, New York, NY 10003, USA

Chapman & Hall Japan, ITP-Japan, Kyowa Building, 3F, 2-2-1, Hirakawacho, Chiyoda-ku, Tokyo 102, Japan

Chapman & Hall Australia, 102 Dodds Street, South Melbourne, Victoria 3205, Australia

Chapman & Hall India, R. Seshadri, 32 Second Main Road, CIT East, Madras 600 035, India

---

First edition 1997

Typeset in 10/12pt Sabon by Acorn Bookwork, Salisbury, Wilts

Printed in Great Britain at The University Press, Cambridge

ISBN 0 412 59410 2

A catalogue record for this book is available from the British Library
Library of Congress Catalog Card Number: 96–72121

∞ Printed on permanent acid-free text paper, manufactured in accordance with ANSI/NISO Z39.48-1992 and ANSI/NISO Z39.48-1984 (Permanence of Paper).

Contents

Foreword

There was a need for a book about the conservation management of fresh waters. It can be argued that there already is a vast literature on the conservation and management of virtually every kind of habitat in almost every part of the world. Moreover, thanks to the computerization of information, this literature, or at least most of it, can be accessed easily and quickly. However, digging out what is wanted from this mountain of information is not so quick and easy. Further, what is accessed also has to be assessed.

When two authors and old colleagues combine to write such a book, we receive a more unified text than is found in the great majority of multi-authored works. Here, we have two authors with long experience of conservation and its management in Britain and elsewhere. In a relatively short handbook they have covered conservation management in such a way that it will be of value to the wide variety of people who and organizations which may play an essential part in one or more aspects of conservation and its management.

Conservation is not simply preservation, like an exhibit in a bottle. It is an active process, which has to continue for an indefinite time. Hence the need for what might be called installation care and after care. The management of both forms a major part of this book.

Conservation management can be an unnatural activity when it holds the natural evolution of a habitat at a certain stage of its development. A common example is the prevention of a habitat from reaching the stage of woodland. Protecting shallow waters, swamps and bogs from further development may be unnatural from the aspect of ecological evolution but what is conserved is a natural habitat with its own specific flora and fauna, part of which is not uncommonly in danger of extinction on a local or even world scale. Many of the diverse aquatic or semi-aquatic habitats, the management of which is described in this book, are indeed endangered habitats: moreover, time is not on our side.

The authors give examples of the conservation management of a wide variety of aquatic or semi-aquatic habitats, ranging from the greatest and oldest lakes in the world to ponds, from great swamps to local marshy areas and outline the skills needed to conserve and manage them. Every case seems to have its own management problems, be it a major political one or a matter of local pressures or prejudices. It is not always a matter of management of a given area; where migrating birds are concerned it may be the management of two or more separate, even widely separate, areas.

This welcome book gives us both the scientific bases for conservation management and much about the art of management of little damaged or even unspoiled, seriously damaged or threatened fresh waters, nor are possibilities of reconstruction of habitats ignored. Along the way, we have interesting examples of conservation management in action – and several horror stories.

John W.G. Lund

1

Introduction

1.1 FRESHWATER HABITATS

Freshwater ecosystems represent a major group of habitats around the world. Not only are the habitats themselves important, for a wide variety of reasons, but the medium itself – fresh water – is of fundamental significance to human welfare everywhere. Fresh water is essential to humans for drinking water, transport, irrigation, energy and as a vehicle to eliminate waste material; the biota provide important food resources and utilizable materials. All freshwater bodies are dynamic systems: not only are their organisms affected by the physicochemical conditions (and thus by human activities), but also the plants and animals interact and may influence both the habitat and one another. They have a major influence on the physical and chemical conditions, while inter- and intraspecific relationships among plants and animals may be of critical importance to both water quality and the structure of communities (Maitland, 1990).

Although the fresh waters of the world seem unimportant compared on an area basis to most land and sea surfaces (only 0.1% of the world's water surface is occupied by fresh water), yet freshwater fishes form 25% of all known vertebrate species. Some of the largest rivers (e.g. the River Amazon, Figure 1.1) and lakes (e.g. Lakes Superior and Tanganyika, Figures 1.2, 1.3) are of impressive dimensions and are of major importance in the general ecology and cycling relationships of the regions in which they occur.

The Amazon discharges annually about one-fifth of the fresh water from rivers entering all the oceans, and is 6500 km in length (Goulding, 1989). Lake Tanganyika has a volume of 18.88 km^3 and is 650 km in length (Coulter, 1991). Such immense water bodies are also of great economic importance to the surrounding regions, both in terms of communication and resources. On the other hand, because of the relatively small size of many wetland habitats, they are very vulnerable to human activities. They cannot be protected completely without control of the whole catchment area. For-

Figure 1.1 The confluence of the Rio Negro with the River Amazon; at this point the river is 7 km wide.

Figure 1.2 Lake Superior: a sandy shore near Marquette, Michigan.

Figure 1.3 The south-east shore of Lake Tanganyika.

tunately, the corollary of this is that, provided the man-induced changes are not irreversible, most freshwater habitats will respond positively to sensible management procedures aimed at restoration, although for exploited peatlands recovery may be extremely slow, depending on the extent of damage.

Freshwater habitats have traditionally been divided into two major types – wetlands (including peatlands) and open waters. They are very variable in character, ranging from running to standing waters, through alkaline marshlands to acid peat bogs, mountain trickles (Figure 1.4) to major rivers and small puddles to lakes over 1.5 km deep (e.g. Lake Baikal, Figure 1.5).

The ecological management techniques necessary for the conservation of habitats are therefore equally variable and the object of this book is to describe some of the principal threats and changes and the methods used to combat them. It is hoped that it contains information which will be useful to a range of people interested in the freshwater environment, from students and naturalists to specialists in the conservation organizations. Before going on to discuss threats and management techniques it is necessary to consider the nature of freshwater ecosystems and the ways in which they function.

1.2 WETLANDS

Wetland habitats are intermediate in character between terrestrial habitats and truly aquatic habitats. They are either permanently wet, as most peat-

Figure 1.4 A mossy mountain stream on Ben Wyvis, Scotland.

lands, or only seasonally wet, such as many marshlands and swamps. They may be transitional areas between open water and the land, as are the fringes of emergent vegetation found along the shores of many lakes. All are dynamic, evolving slowly or rapidly towards a higher level in the seral succession. This evolution may be imperceptibly slow in many peat bogs or relatively rapid in marshes where there is annual accumulation of organic and/or inorganic material, leading to terrestrial conditions. These zones would nowadays be classed as ecotones.

Different authors have varied considerably in what they consider should be included under the term wetlands. The definition of wetlands for the Ramsar Convention (Anon., 1971) is: 'Areas of marsh, fen, peatland or water, whether natural or artificial, permanent or temporary, with water that is static or flowing, fresh, brackish or salt, including areas of marine

Figure 1.5 Lake Baikal, one of the world's largest and most important lakes.

water the depth of which at low tide does not exceed 6 metres'. This definition takes an extremely broad view of what is a wetland and includes virtually every type of fresh and brackish water body, whether static or flowing, plus part of the sea.

On the other hand, the term wetlands is still being used in a much more limited sense, as in the United States and Canada. There the term has been restricted to shallow waters, often intermediate between aquatic and terrestrial ecosystems. Zoltai *et al.* (1975) have defined wetlands as 'areas where wet soils are prevalent, having a water table near or above mineral soils'. For the National Wetlands Inventory Project in the United States, Cowardin *et al.* (1977) have given a similar but more comprehensive definition: 'land where the water table is at, near or above the land surface long enough to promote the formation of hydric soils or to support the growth of hydrophytes. In certain types of wetlands, vegetation is absent as a result of frequent and drastic fluctuations of water-surface levels, wave action, water flow, turbidity or high concentrations of salts or other substances in the water or substrate. Such lands can be recognized by the presence of surface water or saturated substrate at some time during the year and their location within, or adjacent to vegetated wetlands or deep-water habitats.' For inland waters the boundary between wetlands and what were termed 'deep-water habitats' was defined as a depth of 2 metres, or the limit of emergent vegetation if deeper, and for marine systems extreme low water spring tide level.

Thus, in the North American definition, wetlands are characterized by water regime and soil conditions and embrace a continuum from almost terrestrial to truly aquatic systems. Flooding and drying out of the substrate are frequent characteristics. On the other hand the Ramsar definition expands the range of wetlands into the deep water permanent aquatic systems.

The scope of the definitions relates to the purpose of the project concerned. Thus the Ramsar definition covers all the aquatic habitats utilized by waterfowl, the limit of 6 m depth in the sea being the maximum depth to which some waterfowl can dive. The definition of Cowardin *et al.* (1977) meets the needs of the National Wetland Inventory. A division between shallow 'wetlands' and 'open or deep waters' is of course purely arbitrary, as they normally represent an ecological continuum when they are present together. There is exchange of nutrients and energy in both directions, both by water transport and the movement of animals, and the same species of waterfowl may require one habitat for feeding and another for roosting, etc.

Wetlands are so variable in character, throughout different climatic and geological zones of the world, and often merge into other habitats in such a diffuse and variable manner, that it is difficult to arrive at a concise comprehensive definition. A schematic representation of the range of wetland types given by Gopal *et al.* (1990) is shown in Figure 1.6.

For the purpose of this book, 'wetlands' will be restricted to the sense of Cowardin *et al.* (1977), less marine wetlands, and the term 'open waters' will be used for deeper, and generally permanent, standing water bodies, extending below the water depth in which emergent plants will grow, plus running waters. Only natural wetlands will be considered and not man-made wetlands such as rice fields and fishponds.

1.2.1 Peat bogs

Peat bogs are characterized by the accumulation of a peat substrate from undecomposed vegetation under acid, waterlogged and frequently anaerobic conditions. The substrates remain saturated with water, which is extremely poor in nutrients (oligotrophic), for all or most of the year. The plant communities are characterized by the presence of species adapted to these conditions, which supplement their nutrients by nitrogen-fixing bacteria associated with the roots, or by being insectivorous (e.g. sundew, *Drosera*; Venus's fly-trap, *Dionaea*). Peat bogs are formed in areas with moderate to high rainfall and may take many thousands of years to reach maximum development. Eventually the peat level may rise above the water table, so that the bog surface begins to dry out and trees start to establish in a seral succession to forest.

(a) Ombrotrophic bogs

These are bogs which, because of their position, either on mountain plateaux or on plains, are only fed by rainwater, which is their sole supply of

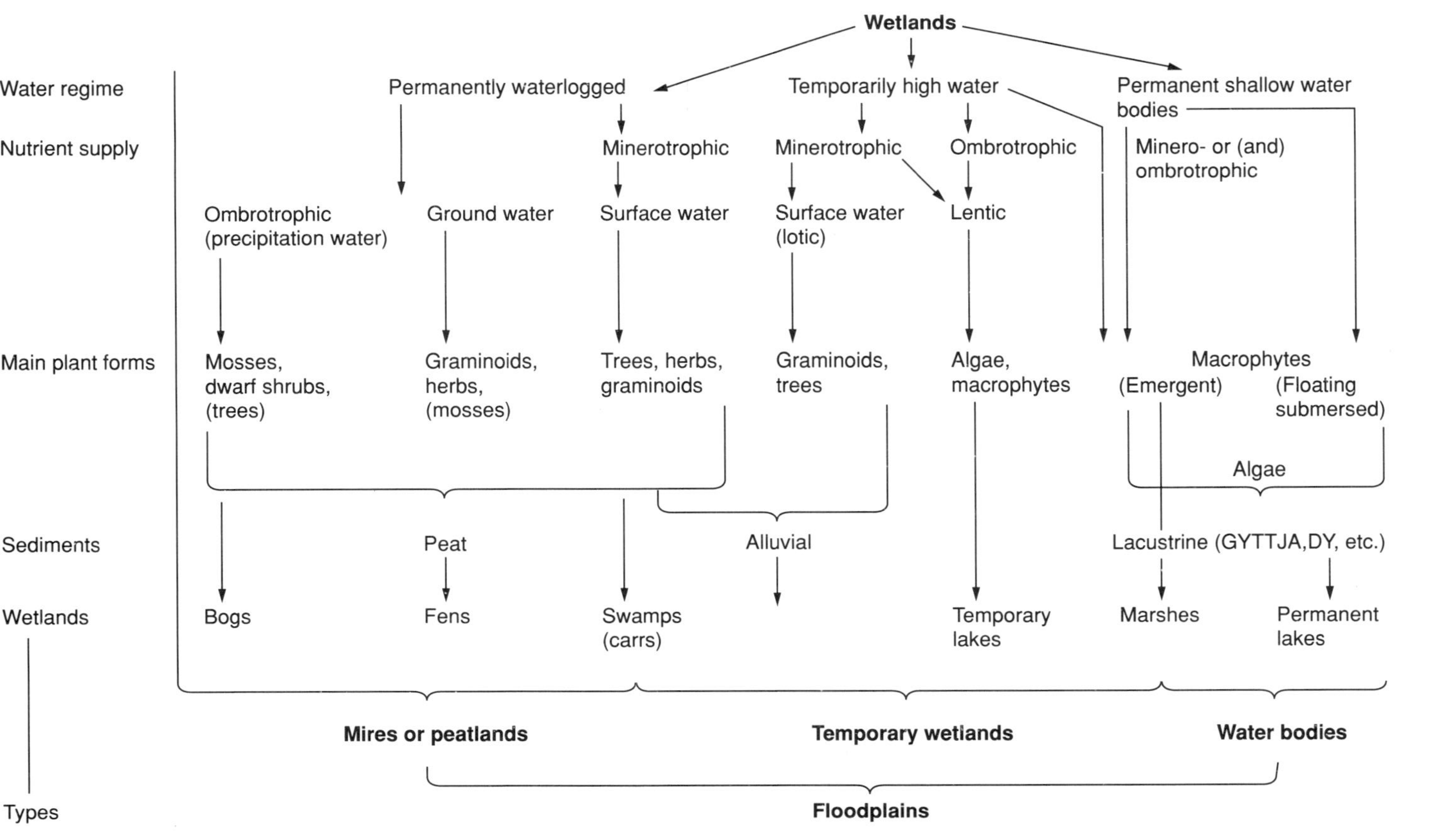

Figure 1.6 Schematic representation of the SCOPE wetland categories and their main parameters and features (after Gopal *et al.*, 1990).

Figure 1.7 A raised peat bog in the Vosges, France.

nutrients. In temperate zones they can only exist where the annual rainfall is more than 160 cm and broadly distributed throughout the year. Where the rainfall is sufficient, the vertical growth of peat can progress above the water table so that a dome-shaped 'raised bog' is formed (Figure 1.7).

The vegetation of raised bogs is dominated by *Sphagnum* species. The paucity of nutrients, particularly nitrogen, in many peatlands – especially those that depend entirely on direct rainfall – has led to specializations to obtain the necessary nitrogen. Insectivorous plants compensate by digesting insects, while certain blue-green algae and bacteria fix atmospheric nitrogen. Some bacteria often form commensal relationships with higher plants.

(b) Soligenous bogs

These are peatlands that receive their water principally by runoff from the surrounding land (Figure 1.8).

Although low in nutrients they are richer than the ombrogenous bogs, but they remain acid. *Sphagnum* mosses are less prevalent than in ombrogenous bogs but species with more exacting demands, e.g. *Eriophorum, Trichophorum caespitosa* and *Molinia caerulea*, become important. The latter may completely dominate in some blanket bogs. Soligenous bogs have been subdivided into various types such as blanket bogs, parts of which may be ombrotrophic, valley bogs, floating bogs (Figure 1.9), etc., based on physical and botanical differences.

Figure 1.8 A soligenous bog in northern Finland.

Figure 1.9 A floating bog in the Vosges, France.

Figure 1.10 The Djoudj marshlands, Senegal, with *Nymphaea caerulea* and *Actophilornis africana*.

1.2.2 Marshlands

(a) Marshes

These are wetlands dominated by herbaceous vegetation that may be seasonally or permanently flooded. They may be fed with water from vertical movements of the water table or from inflowing streams, or may occupy the littoral of lakes from which they are flooded. Their water is neutral to alkaline, and the nutrient level moderate to rich. There is little or no accumulation of peat and the substrate is usually mineral or a mixture of mineral and organic material. *Phragmites* and *Typha* are cosmopolitan marshland plants and *Cyperus* and *Nymphaea* typical in the tropics (Figure 1.10). With changing substrate or water level marshes may progress into swamps.

(b) Fens

These are intermediate in character between peat bogs and marshes. They are less acid than neighbouring bogs and, although the substrate is composed of peat, this is usually in a moderate state of decomposition. The water supply is generally from ground water and precipitation, and dissolved oxygen levels are relatively low. The vegetation is dominated by *Carex* species. *Sphagnum* is reduced or absent and its place is taken by other mosses.

Figure 1.11 Horseshoe Lake, Illinois, with swamp cypress, *Taxodium distichum*, and black gum, *Nyssa sylvatica*.

(c) *Swamps (carrs)*

Swamps have a wide range of vegetation types ranging from mosses, grasses, herbs and shrubs to trees. Whatever the variation in the composition of the understory they are always dominated by trees adapted to inundation, often with special root structures, such as pneumatophores. These trees are generally broad-leaved species but some conifers, like the swamp cypress, *Taxodium*, are adapted to these conditions (Figure 1.11).

The water is generally neutral to moderately acid and *Sphagnum* may be abundant in the latter conditions. The substrate is variable, from inorganic through decomposing organic to peat, and is often continually waterlogged.

(d) *Floodplains*

These are areas of low-lying flat ground over which rivers flood during high water. In regions with distinct wet and dry seasons, as in Africa, flooding is confined to the wet season and is predictable (e.g. Okavango and Bangweulu wetlands), but in temperate zones, where heavy rainfall is unpredictable, flooding can occur at any time of year. In many areas the flood plains have been modified for farming, e.g. pasture, hay or rice, but in others, such as Amazonia, large tracts of riverine forest remain (Figure 1.12).

The tolerance of different species of tree to the height, frequency and duration of flooding varies, and the composition of alluvial forests varies in

Figure 1.12 The flooded Amazon forest.

relation to this, within any one climatic zone (Yon and Tendron, 1981). The flooded forest of Amazonia is well over 100 000 km^2 in area and inundation of the trees may last as long as 11 months, depending on the height of flooding (Goulding, 1989).

Elements of all the wetland types described above are found associated with floodplains. The quality of the water is dependent on the inflowing rivers and the substrate contains alluvial material deposited by them.

1.3 OPEN WATERS

Open waters neatly classify into standing waters (ponds and lakes) and running waters (streams and rivers). Though occasional problems may arise (e.g. with canals and the backwaters of slow-flowing rivers) there is usually little difficulty in deciding whether a water belongs to the standing or the running water series. Rarely are the two unconnected, however: running waters often have one or more standing waters along their length; and most standing waters receive several running waters, and water exits via a single running outflow (Figure 1.13).

Functionally and ecologically, running and standing waters may be differentiated as follows:

- A unidirectional current (created by gravity) is found in running waters, but currents (mostly due to wind) are variable and weaker in standing waters.

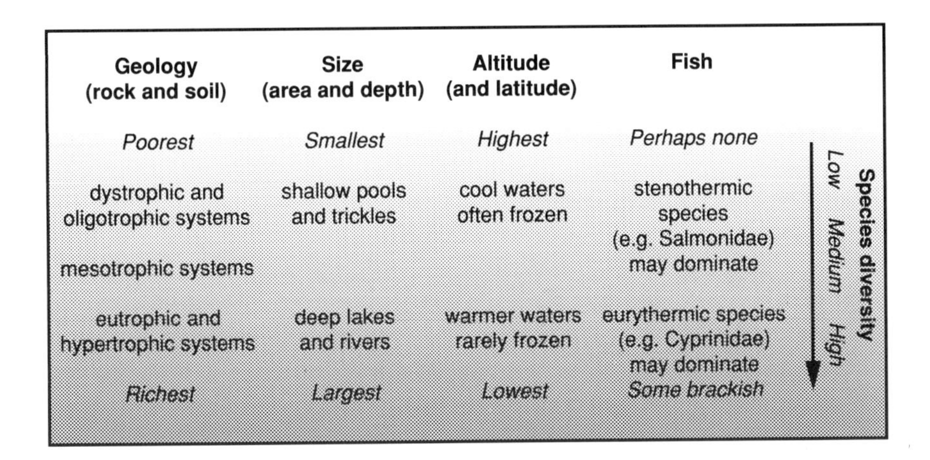

Figure 1.13 A simple diagrammatic classification of freshwater habitats. The diagram represents a continuum, which is actually the commonest situation, for most water bodies interconnect with others. In some large catchments all these different water types may be found.

- Running waters rarely stratify, whereas stratification is a characteristic feature of many standing waters.
- Longitudinally, physicochemical conditions in running waters change gradually but the difference in many parameters may be great from source to mouth. Conditions in standing waters are normally much more homogeneous.
- Running waters are always relatively shallow and have long, often complex, narrow channels (Figure 1.14). Standing waters may reach great depths, but most have simple, often broad, basins (Figure 1.15).
- Erosion is characteristic of running waters, and materials so removed may be transported considerable distances, often outwith the catchment concerned. Strongest erosion takes place in the precipitous mountain reaches of rivers and much of this material is deposited in the slow meandering reaches or on the flood plain. Erosion does occur in standing waters, but is only severe in long exposed reaches and eroded materials usually remain within the same basin.
- As a further consequence of erosion and deposition, most running waters increase the length of their channels with age, as cutting back to the source and meandering on the flood plain proceed; in standing waters, materials are constantly being deposited, tending to fill in the basin and eventually obliterate it completely.

Thus the ultimate fate of many standing waters (especially shallow ones) and parts of the channels of some running waters (e.g. when ox-bow lakes

Figure 1.14 The River Wye, Wales, an example of diverse habitats in a short stretch of river.

Figure 1.15 Bangula Lagoon, Malawi, a major resource for local people.

are created) is to become marshlands as silting and ecological succession from open water progresses.

1.3.1 Physical environment

In most fresh waters the major source of energy is solar radiation and thus the amount of radiation entering a water body is of vital importance to the way in which it functions. The total amount of radiation reaching the surface of a water depends on time of year, geographic location, altitude, state of the atmosphere and several other (usually local) factors. As well as being reflected from the surface, some radiation is scattered from below the water into the atmosphere. In most waters the amounts involved in each loss are about the same.

Thus waters may differ enormously in the amount of energy which they receive from the sun. Not only are waters at widely differing latitudes different in this respect, but almost adjacent waters may be too. Though two lakes may have the same volume, if one is wide and shallow and the other is narrow and deep the former will receive much more energy than the latter. Shading, too, whether it be from mountains or from large trees, can substantially lower the solar input to an aquatic system. Some of these factors can be managed but others cannot.

In observing natural fresh waters, the colour may be due to an actual colour, or an apparent colour made up from this and the influence of other factors. The quality of the incident light, the selective transmission of wavelengths, the amount and quality of suspended matter and (in shallow water) the colour of the substrate are all important. Basically, the colour observed on looking into a river or lake is that of the upward scattered light. Natural systems where the water is relatively pure appear very dark or bluish (Figure 1.16), unless they are very shallow (where the colour of the substrate is important) or reflect colour from their surroundings. In waters where there are large quantities of suspended materials, either living (e.g. algae) or inanimate (e.g. clay), coloured light is reflected and combines with the transmission effect to give a particular colour. In natural waters where there are small amounts of suspended matter the water looks green but where there are large amounts the water looks yellow or brown.

Monitoring the colour of a water body on a regular basis can prove a useful long-term strategy to detect (and if necessary attempt to reverse) important changes, such as those described in section 3.1.2.

The presence of ice can be a major factor in controlling the amount of radiation entering water bodies in temperate and arctic areas. The transparency of absolutely pure ice is very high and with such 'black' ice, a relatively high proportion of the solar radiation penetrates below. Where, however, the ice is less transparent (usually where it has been formed under

Figure 1.16 Loch Morar, Scotland, one of the clearest and deepest lakes in Europe.

rough conditions) or covered by snow or dirt, very little energy may reach the water underneath because of loss by reflection or absorption.

The specific heat of water is high and indeed much higher than that of many other materials. This property means that temperature conditions in water are much more stable than those in air: rapid diurnal and seasonal changes of temperature in natural waters are rare – an important factor in the ecology of many aquatic organisms.

Water is critical to life on earth, which is unusual among the planets in having large amounts of water on its surface. Water itself is remarkable in being almost the only major material existing as a liquid on the earth's surface at ordinary pressures. At 0°C and atmospheric pressure (760 mm Hg) the density of pure water is more than 700 times that of air. This means that the tissues of aquatic organisms need far less support than those of terrestrial ones, and there can be a great reduction in skeletal structure – this is especially valuable for large animals. The density of water in different places can vary in time and space, and even small variations are important. They are mainly due to temperature and dissolved solids, especially the former.

The density of most materials increases with decreasing temperature, whereas, remarkably, water reaches a maximum density at about 4°C. This anomalous change in density is of considerable significance biologically (and to many management practices) because, after cooling, water starts freezing at the surface but near the bottom the temperature may be greater – usually about 4°C. The actual formation of ice cover depends on a variety of factors. The most critical of these are that the water temperature in the whole water body is 4°C or below and that the weather is clear with little wind. Conditions for the break-up of ice cover are the opposite of these.

Ice cover can be a critical factor in controlling the nature of a water body and the character of its plants and animals. For example, in waters with high amounts of organic material on the bottom (e.g. aquatic vegetation or tree leaves), oxygen may be used up rapidly under ice and 'winter kill' results. Such waters have no fish and only specialized invertebrates (such as those with haemoglobin) that withstand such conditions.

Temperature is an important factor influencing the biology of fresh waters. Most freshwater animals and all plants are poikilothermic, and thus their temperature varies with that of their surroundings. The physiological reactions taking place within them (e.g. photosynthesis, respiration, digestion) are biochemical ones and the rate at which they take place is very dependent on the ambient temperature. Many freshwater organisms react to temperature with considerable precision and are capable of detecting differences as small as 0.2°C in their surroundings. The life cycles of temperate organisms are geared to the seasons: the breeding biology of many insects is closely linked with the day-length and/or temperature of their surroundings.

Summer temperatures may be critical for many cold-blooded vertebrates, which require to grow rapidly enough to be able to withstand winter conditions. This is true of many cyprinids and amphibians. Some invertebrates too require warm summers for development: for example the endangered medicinal leech *Hirudo medicinalis* is only successful in the northern parts of its range in shallow sheltered ponds that reach over 20°C during summer (Elliott and Tullet, 1992).

The nature and extent of water movements is influenced by several factors. Because of the importance of winds in producing movements in standing waters, and gradients in causing currents in running waters, local topography is of considerable significance. The shape of the basin or channel is also important. In very large bodies of open water, currents may be influenced by the rotation of the earth, as is the sea. In open waters the energy producing most currents is derived from the wind (the dominant force in lake currents), from changes in density within the water, or from kinetic differences in level (the major factor in determining currents in rivers).

Currents often define the nature of the substrate in a water body and are thus one of the most important factors affecting its ecology. Therefore, control of water movement, where this is possible, is often a very important management tool.

Figure 1.17 The Dead Sea: high quantities of dissolved salts aid buoyancy!

1.3.2 Chemical environment

Although pure water is H_2O, fresh water itself is normally far from pure. Almost all natural waters, even rain, contain various chemicals, though the concentrations of different substances may vary greatly from one water to another (Figure 1.17).

Most waters, too (apart from those which are toxic or of very high temperature), possess living organisms, which create a dynamic system, so that the chemical conditions are constantly changing. An understanding of the chemistry of any water body is an important prerequisite to successful scientific management.

Five main environmental factors interact to determine the ionic concentration of natural waters: climate, geology, topography, biota and time

(Gorham, 1961). The chemicals occurring in fresh waters originate not only from the soils and rocks of the catchment but also from the atmosphere. Materials in the atmosphere, which come from air pollution (especially combustion) by humans and volcanoes as well as from wind-blown sea spray or land dust, enter the system with rain or snow, fall out dry or are transferred as gas. Substances from the earth, which are normally more important than those from the atmosphere, are dissolved or released by various chemical reactions. The transfer from soil or rock to water is influenced by ion exchange and the type of water involved.

In oceanic areas the total amounts of salts deposited in rainwater are high compared to continental areas and characterized by a high proportion of cyclic salts (Holden, 1966). The proportions of sodium, magnesium and chlorides in precipitation are similar to those in seawater, but the proportions of potassium, and especially calcium and sulphate, are in excess. Much of the latter is supposed to be derived from industrial activity.

Thus, in natural fresh waters, the concentrations of dissolved solids vary greatly but an average value is approximately 100 mg/l. Water with less than 50 mg/l indicates drainage from igneous rocks, while water with more than 100 mg/l points to drainage from sedimentary rocks. As the amounts of dissolved salts increase, the proportions of calcium and magnesium tend to rise, and those of alkalies to fall. Where large areas of highly organic peaty sediments occur, very soft water with a low concentration of calcium but a relatively high proportion of alkalies may occur.

Oxygen and carbon dioxide are of major importance in aquatic systems, and where variations in concentrations occur they normally show an inverse relationship. The respiration processes of plants and animals and the oxidation processes of breakdown lower the amount of oxygen and increase the carbon dioxide. With sufficient light, the process of photosynthesis by living plants produces the opposite effect – an increase in the amount of oxygen and decrease in carbon dioxide.

A further factor of major significance to aquatic systems is the amount of movement and mixing of the water, especially its contact with the atmosphere. If adequate mixing is prevented, e.g. by ice, extreme concentrations may occur, i.e. supersaturation where thick plant growths are exposed to bright light under ice, or anaerobic conditions where there are large numbers of animals or much decomposing organic matter and no light for photosynthesis. Limnologists have traditionally used the demand for oxygen by oxidizable organic matter as a means of comparing different waters, especially those affected by organic pollution. The biological oxygen demand (BOD) of such waters is a purely arbitrary measure of the amount of oxygen used by a sample of water during a standard period of 5 days at 20°C in darkness.

Different natural waters exhibit widely differing values of pH and records range from below 2.0 in certain volcanic lakes that contain free sulphuric acid to more than 12.0 in highly alkaline soda lakes. Apart from bacteria, few organisms can tolerate such conditions.

Carbon dioxide is absorbed by rain falling through the atmosphere and combines with water to form carbonic acid, H_2CO_3. In areas of limestone or other calcium-bearing rocks this acid acts on the rocks to release a soluble product – in the case of limestone, calcium bicarbonate, $Ca(HCO_3)_2$. Water normally emerges as a spring or trickle of some kind after passing through the soil. Calciferous waters then start to lose carbon dioxide to the atmosphere or, on mixing with other water, the calcium bicarbonate dissociates and insoluble calcium carbonate precipitates, often encrusting the substrate and forming marl.

If calciferous water comes into contact with acid water, some bicarbonate combines with acid to release free carbon dioxide, which is only weakly dissociated, altering the pH by a small amount. Only small changes in pH can occur until all the bicarbonate is used up. This buffering effect prevents drastic changes in reaction and is important to aquatic organisms. In most natural waters the buffering ability is determined by the amount of bicarbonate present; this and the pH are useful parameters to measure when investigating their chemistry. Waters that are poorly buffered may exhibit drastic fluctuations in pH and can be alternately acid and alkaline on a circadian or some other basis, depending on local circumstances.

The amounts of nitrogen available in a water, though often small, are of significance in most ecosystems, for nitrogen is an important component of the cells of living organisms. In the majority of natural aerobic waters, most nitrogen occurs as nitrate. In eutrophic waters with large standing crops of algae or macrophytes, almost all the nitrate present may have been incorporated in the plant cells. In anaerobic situations nitrates are often broken down through nitrite and nitrous oxide, occasionally as far as nitrogen itself. In addition, in the absence of dissolved oxygen, free ammonia may occur as a result of protein breakdown.

The quantities of phosphorus (occurring mainly as phosphate) in most waters are low even though this element is an important constituent of living organisms and present in them in significant amounts. This is because the element is naturally scarce and because of the capacity of many plants to absorb and store many times their immediate needs of phosphate. Both phytoplankton and littoral vegetation are capable of doing this; moreover, many of these plants are exceptionally efficient in using phosphorus and can extract it in significant amounts from water where it is present in minute quantities. This property is exploited in modern biological techniques for the purification of organic sewage. Rainwater normally contains very little phosphorus but when the rain reaches the earth and percolates through the soil small amounts are leached from phosphorus-containing rocks and from phosphorus present in the soil. The quantities occurring naturally in fresh waters depend on the geochemistry of the catchment, and are greater in areas of sedimentary rocks than in those containing igneous rocks.

Iron is involved in the cycling of certain nutrients (e.g. phosphorus) and plays an important part in the metabolism of many organisms. Iron normally exists in two forms: ferrous, which is soluble but stable only in anaerobic conditions, and ferric, which is stable in the presence of oxygen but is insoluble. Thus in most natural waters there is very little dissolved iron.

Many other chemicals are of importance to the ecology of wetland systems. The role of calcium has been mentioned in connection with pH and carbon dioxide. Magnesium, similar in many respects to calcium, is often present in solution as bicarbonate. Its monocarbonate is much more soluble than that of calcium. Silica forms an important part of the skeletal structure of diatomaceous algae and some animals. Occurring normally as undissociated orthosilicate, most is derived from silicious rocks in the catchment. Sodium and chloride ions occur in most waters, but in low quantities unless the system is affected by saline ground water or ocean spray. Both ions are of major importance in brackish and salt waters. As in all ecosystems, trace elements are important in fresh waters; relatively little, however, is yet known about their true role there.

Once thought to be of little importance in aquatic systems, dissolved organic matter is now known to be of considerable significance and may occur in far greater quantities than suspended materials in many waters. Though important quantitatively as a source of carbon, the quality of the organic material can be of even greater significance, for it may contain valuable amino acids and vitamins. Such materials may originate within a water through the growth and reproduction of organisms; substances derived in this way are termed autochthonous. Materials which originate outside the aquatic system and are subsequently washed or blown in are known as allochthonous.

It will be clear from the above that the chemistry of any water is a major factor controlling its ecology and that if its chemistry can be managed then an important tool is available to the conservationist. This topic is discussed further in sections 7.3.3 and 8.4.3 but it is worth mentioning here that 'point' sources of input to a water (i.e. specific discharges, tributaries, etc.) can be managed relatively easily, whereas 'non-point' or diffuse sources (i.e. those entering widely across the catchment, such as agricultural drainage) are usually very difficult to control.

1.3.3 Biological environment

Primary production by plants is the basis of almost all food chains in fresh waters and thus control of vegetation is a key factor in conservation management. Photosynthesis, one of the major life processes occurring on earth, is fundamental as the means of carbon fixation. The gas exchange system related to photosynthesis is rarely in constant equilibrium but usually exhibits a circadian cycle on which a seasonal pattern may be imposed. During darkness, when oxygen is used continuously for respiration by plants, animals and breakdown processes, dissolved oxygen content in the water decreases; it is replaced only by mixing processes at the water surface. During darkness, carbon dioxide is produced continuously; it is either used to build up a store of dissolved bicarbonate or lost at the water surface. With daylight, however, assuming that adequate solar radiation is available, photosynthesis proceeds and oxygen is given off. This may counteract any deficit that has occurred during darkness and, depending on the rate of circulation of the water and the degree of gas exchange at the surface, may even build up to supersaturation values.

The requirement of plants for light in order to photosynthesize means that the depth to which any species can grow is determined by the extent to which sufficient light for its requirements can penetrate. Different waters may have widely varying light extinction coefficients depending on the quantity and quality of dissolved and suspended material (including algae) that they contain. Generally, there is inadequate light, even in very clear waters, for plants to grow at depths greater than 10 m, and in many eutrophic or turbid waters photosynthesis cannot proceed below about 2 m.

Successful growth and reproduction of plants in fresh waters requires the presence of several nutrients. These are present in very variable quantities from one water to another and even within the same water system may vary in space (especially during stratification) or seasonally according to rainfall, temperature, wind and the nature of the plant community itself. Usually, several nutrients are present in excess of what might be required for the adequate growth of a particular plant species. If even one important nutrient is scarce, it may severely hinder the production of a species and prevent it using other nutrients present in excess. Nutrients known to be limiting in natural waters are carbon, nitrogen and phosphorus, but the requirements vary with different plant species and with each community.

Nitrogen can be a major factor limiting primary production in eutrophic and some tropical waters, but in many waters phosphorus occurs only in minute amounts and is an important controlling factor. Magnesium, potassium, iron, manganese, cobalt, molybdenum and zinc may also be of importance in limiting growth in various waters.

It is clear that living organisms are profoundly affected by their environment. It is equally important to remember that many organisms are capable of doing the reverse and significantly altering their habitat, sometimes to their own detriment. The influence of the biological component is often relatively greater in fresh water than in marine or terrestrial systems because of the small size of many waters. There are many examples of the influence of organisms on fresh waters and only a few cases are considered here.

Photosynthesis is a notable example of a major influence of plants on their environment and by their subsequent growth many species can deplete essential nutrients, thus limiting their own growth or that of other species. In Windermere the alga *Asterionella* is unable to grow in conditions which it itself has created (Lund, 1950). This plant starts to grow rapidly in the spring in this lake, using up so much of the silica that there is no longer enough to maintain its own growth. The population declines as a result. If an algal population is not limited by nutrients, its rate of growth may be rapid but, when certain densities are attained, self-shading may occur and control the population at this level.

Large amounts of oxygen may be used up by the respiration of animals. Edwards (1958) has shown that the burrowing activities of the larvae of the midge *Chironomus riparius* considerably alter the character of the mud in which they live by extending the redox potential further down into the mud. These larvae, when added to natural muds at densities of about $8000/m^2$, raised the oxygen uptake some 25% – equivalent to their predicted respiration. Edwards estimated that natural populations of this species are capable of lowering the oxygen concentration of a stream by almost 1 mg/l. Such a process can be taken to extremes where there is no replacement of the oxygen, as in the hypolimnion or under ice in some lakes. Here, oxygen deficiency can cause massive mortalities among the invertebrates and fish present.

Plants and animals, as well as altering the chemistry of their environment, often affect its physical characteristics profoundly. The relationship between plants and light extinction has already been discussed. Several species of macrophyte form such large growths in rivers that they seriously affect water levels and velocities. Hillebrand (1950) has shown that in the River Eder, in a stretch where the breadth of the river was some 70 m, the dense weed grew so quickly in summer that the water level rose by over 1 cm per day. Such growths may also change the character of the substrate for, though originally stony or gravelly, with the development of macrophytes, water movement is impeded and increased siltation results. In this way the original substrate may be completely obscured by the new sediment. In lakes too the growth of macrophytes may have a profound effect and, through the process of ecological succession, small open waters may be transformed into marshes and eventually to dry land. Similarly, as plant growth in ombrotrophic peatlands gradually raises the surface above the water table, raised bogs start to become senescent and dry out – with invasion by trees, such as *Betula* and *Pinus* spp. – and are eventually transformed into woodland. Completely new lakes, on the other hand, can be created by certain organisms, the dams constructed by beavers being a notable example of this.

Clearly, therefore, an understanding of the ways that aquatic plants and animals interact with each other and with their environment is important if their management is going to be attempted on a scientific basis.

2

The value of fresh waters

2.1 DEMANDS FOR FRESH WATER

Fresh water has been used by humans from earliest times, at first only for drinking, but later for fishing and navigation. The majority of settlements in many countries are related to spring lines and other sources of pure water. With improving sanitation, water was used for cleaning and removing domestic wastes and for irrigation in agriculture, and its power was harnessed for driving industrial machines. Within the last two centuries, improving standards of living, increased sophistication of agricultural methods, industrial development and production of hydro-power have meant that water has become more and more important to humans. Further, extended leisure time in modern societies has increased pressure on recreational facilities for aquatic activities like angling, wildfowling, sailing, swimming, water skiing and power-boating.

Because fresh water is essential to successful modern societies, a useful way to compare living standards of different nations is to examine the *per capita* consumption of water. Countries with a high standard of living (e.g. Norway, Sweden and Switzerland) show a high requirement for clean fresh water (*ca* 100–500 litres per head per day) compared to the poorest countries (less than 10 litres). The ever-increasing standards of living in most countries and the increase in human population combine to make heavy demands on freshwater resources, and result in conflicting interests in available water. Production of fresh water from the sea by desalination, and recycling of water after treatment by sewage works are among the approaches being adopted to combat this problem.

The quantity of water required for domestic and industrial (including agricultural) purposes has shown no sign of lessening in recent years, making it necessary to consider the whole question of water resource and supply on a more extended basis. Water conservation means the preservation, control and development of water resources (by storage, prevention of pollution or other means) to ensure that adequate and reliable supplies are available for all purposes in the most suitable and economic way while safeguarding legitimate interests. Originally, only extremely pure waters were considered as potential sources, and this emphasis on quality led to excessive exploitation of ground water and impoundments in upland areas. The increasing shortage of water, and the improving means for treatment of contaminated water, have meant that quantity and not quality is often more important; sources which would not have been considered in the past are now accepted and treated for public supply. Excessive use of water is now controlled by metering and charging users in relation to the quantities used.

In areas of low rainfall around the world, particularly those far from large natural supplies, the problem of supplying large volumes of suitable water may be expensive or insoluble. Even in areas of adequate rainfall, seasonal variation may cause difficulties, low rainfall creating a shortage, high rainfall causing flooding. This situation is typical of many countries, where the remedy is one of national expenditure on the development and conservation of water resources. The provision of water for domestic stock is important and in desert areas stock distribution is limited by the presence or absence of sources of fresh water.

2.2 MANIPULATING FRESH WATER

2.2.1 Sources

Precipitation is, of course, the initial source of water, and a study of the rainfall and hydrological cycle is a necessary preliminary to assessing the water supply potential of any area. Only a relatively small proportion of the total rainfall in a large geographic area is readily available for water supply; major losses occur from evaporation during precipitation, from the ground or open water and from transpiration from vegetation. Enormous amounts of water flow into the sea directly. Utilization of this available rainfall involves collecting it as surface water (by intakes or pumping from rivers, or by piping from suitable lakes or reservoirs) or drawing on ground water (by springs or by sinking wells).

Mountainous regions, where natural systems are oligotrophic, are the most suitable areas for using existing lakes or establishing reservoirs. Waters in highland areas are especially suitable for domestic supplies in that there is less pollution than in lowlands. Also, the initial rainfall is higher in highland than lowland areas. Oligotrophic waters contain little suspended matter (especially algae) and require little filtration. Though the geographic regions most suitable for water supply bodies are often far from areas where water is most needed, the altitude of such systems means that water will readily pass by gravity, obviating the need for expensive pumping. As well as the chemical nature of the catchment its physical geology (especially where man-made reservoirs are concerned) is important. Local geology affects the amount of direct run-off, the potential underground losses and the mechanics and cost of dam construction.

Ground water collects over impermeable strata, where it may remain for long periods, after it has fallen on the earth as rainfall and then percolated through soil and rocks. The surface of this ground water within the earth (comparable to the surface of a lake) is known as the water table and may fluctuate according to losses from evaporation and underground run-off or additions from rainfall. Other than where large springs appear at the surface, the collection of ground water involves the sinking of a well and then pumping water which drains into this from the surrounding pervious strata (the aquifer). Under natural conditions, ground water, which is filtered in passing through the ground, is usually highly acceptable as drinking water. In areas with intensive agriculture, the application of artificial fertilizers has led to the building up of levels of nitrates in the water table which, in some areas, are now a danger to human health.

Less desirable as sources of potable fresh water than either highland water or ground water, river water is often used in lowland areas. Normally, pumping is necessary and the water often carries considerable amounts of suspended solids and may be polluted and require considerable treatment. Nevertheless, in some areas rivers are becoming the most important source of supply. The intake pipes are strategically placed so that they will always be under water but will be little affected by strong currents or silting. It is common practice to install a weir to raise the river level for abstraction and to deflect the water (especially at low flows) into the intake.

2.2.2 Storage

In areas with low or variable precipitation it is essential to construct adequate reservoirs to make economic use of water resources. These form a component in most water supply systems and in many areas there are now more reservoirs than natural lakes. The recent construction of very large systems like the Kariba, Aswan and Bratz Dams (Figure 2.1) makes these reservoirs among the largest freshwater bodies in the world.

For many purposes (irrigation, some industries, navigation and power production) the quality of water is not of primary importance, the main factors being the amount and head available and the efficiency of the system in relation to catchment hydrology. With reservoirs for domestic supplies, on the other hand, it is essential to consider not only the physical and engineering aspects of the system but also the biological ones.

An initial settling basin within a reservoir system is often desirable to sediment incoming silt. Main reservoir basins may be shallow or deep; if shallow, there is less likelihood of stratification but there may be a higher level of algal production. In deep reservoirs, algal production will be lower

Figure 2.1 The giant Bratz Dam, Russia.

(because less water is exposed to light) but stratification is probable and may lead to anaerobic conditions in deep water. The resulting odour and taste may make water here unsuitable for immediate domestic supply.

2.2.3 Flood control

Shortages in quantity or quality are the most common difficulties met in providing water. With flooding, however, excess is the problem and one that causes tremendous damage to life and property in many parts of the world; in a number of areas, massive and expensive flood prevention schemes have been installed. Many problems connected with flooding arise from the fact that the most desirable industrial, agricultural and residential areas lie on flood plains close to rivers and that increase in river canalization and land drainage schemes causes water falling on uplands to find its way to the flood plain much more quickly than under natural conditions and to exceed the capacity of the floodplain to accept it.

Flooding in lowland areas can be prevented in several ways. River-training works (e.g. stone or concrete walls, willow piling or groynes) can be constructed to give permanency to the river channel and prevent damage by erosion. The area flooded may be reduced by constructing flood defences (earth embankments or concrete walls). The water level in the main river can be reduced by providing flood relief channels, by enlarging the main channel (either with embankments or by dredging and widening), or by intercepting the flood water before it reaches the danger area and diverting it through a new intercepting channel or into a flood storage reservoir. In some cases pumps are used to deal with urban flood problems. Many modern schemes incorporate several of these methods. Such interventions are, however, only palliative and often lead to excessive discharges downstream. They are no substitute for the storage and gradual discharge of water in natural floodplains and other wetlands along the river course.

2.3 USING FRESH WATER

2.3.1 Domestic needs

The provision of wholesome water is a vital element in the health care of modern communities. Organizations supplying water for human consumption are concerned with producing as economically as possible adequate volumes that are free from objectionable odour and taste. This water must also be clear and free from harmful mineral substances, as well as disease organisms (e.g. certain bacteria and viruses). Among the important human diseases spread by water are bilharzia, amoebic dysentery, gastroenteritis, leptospirosis, infectious hepatitis, cholera and typhoid. Large amounts of some inorganic salts may cause dental fluorosis (excess of fluoride), methaemoglobinaemia in infants (excess of nitrate) and poisoning due to lead or other heavy metals. Adequate precautions to prevent pollution of the water supply and efficient treatment before consumption would be sufficient to avoid troubles from most of these causes.

Throughout the world, many reservoirs have been constructed for storage, so that water produced by high rainfall may be available for supply during dry periods. In some cases, however, especially with river waters, storage in large reservoirs is a preliminary form of treatment, when the water loses much of its suspended matter and harmful organisms are destroyed. Colour and some forms of nitrogen may also be reduced, but storage can involve an increase in the algae and other plankton present. Most waters for domestic supply require further treatment to remove suspended and dissolved solids, colour, taste and odour. This may involve sieving, coagulation, sedimentation, filtration, sterilization and softening.

2.3.2 Agriculture

Agriculture uses large amounts of water (from open water and ground water), by far the greatest percentage for the growth of crops, much of this being immediately passed into the atmosphere by transpiration. In developed countries, a small percentage is used for consumption by stock, vegetable washing and implement cleaning. In some countries adequate rainfall meets the needs of outdoor crops, but in many others, including some in temperate areas, precipitation is inadequate and supplementary water must be provided by artificial irrigation. In arid areas, agriculture is possible only if adequate water for irrigation is available and the key to the development of many countries lies in providing water for this purpose at an economic cost.

In several countries the requirements for irrigation are more seasonal than those for domestic water supply, and unfortunately tend to be greatest when rainfall and artificial storage reserves are lowest. Specific irrigation requirements vary according to local conditions and the type of crop grown, and must normally be calculated for each area (Prickett, 1963). Irrigation is worthwhile only where the costs of application over a period of years are covered by the increased return from the crops so grown. This is often possible by only irrigating small areas on which the intensive culture of high-value crops is practised.

2.3.3 Aquaculture

Though many of the techniques used in modern aquaculture are highly sophisticated, the culture of fish and other aquatic organisms (e.g. shrimps, crayfish and other Crustacea) is a very ancient form of rearing animals for food and has been carried out in China for more than 4000 years. In many areas the methodology and systems involved have changed little over hundreds of years. In Europe, the records do not go back further than the 12th century, since when many species of fish have been widely kept. In Great Britain, many of the early waters were stew ponds associated with monasteries, but in recent years aquaculture has expanded rapidly. Elsewhere in Europe, in Russia, the Middle and Far East (especially China), some parts of North America and many areas of the tropics, the rearing of fish and crustaceans is an important and expanding part of the economy, particularly as far as the production of protein is concerned.

Until recently, aquaculture was mainly carried out in fresh waters, or occasionally in brackish waters in some tropical areas, but marine aquaculture is now important in some parts of the world. With some species (e.g. Atlantic salmon *Salmo salar*), both fresh and salt water are needed to complete the life cycle (Figure 2.2).

With efficient methods and suitable local conditions, extremely high rates of production are possible compared to other forms of protein production. Moreover, aquaculture is often the only practicable use for some types of ground – where, for example, drainage is poor or ground water too saline for traditional agriculture. Fish farming is often compatible with other farming: in Africa and Asia, fish, ducks and cereals are grown together in irrigation schemes while in Europe many fish ponds are dried out in alternate years to grow cereals.

Figure 2.2 A modern cage farm for Atlantic salmon, Mull, Scotland.

Successful aquaculture requires suitable water, an adequate area for ponds (with suitable soil), local supplies of fertilizers or food, adequate labour and experience to operate the farm and a good local market for the end product, or freezing facilities and transport to such a market. The water required depends on the type of aquatic organism being kept: salmonid fish require large amounts of cool, high-quality water while carp and many crustaceans will tolerate water of poorer quality without a rapid flow-through of water. With adequate water, a site is required where the soil will retain it or can be made to do so relatively easily. It is always advantageous if ponds can be drained and rarely necessary to have them very deep – small ponds for young fish are rarely deeper than 1 m and very few artificial fish ponds are more than 2 m in depth.

The crops produced by aquaculture can be variable and depend very much on the species of fish, the management techniques and the amount of energy put into the system, whether in the form of solar radiation or direct feeding. All ponds have a maximum standing crop; this is not necessarily the best one for maximum production. The production potential is controlled by the maximum standing crop and the time taken by the initial stock of fish or crustaceans to reach this level. The natural standing crop of a pond may be increased several times by fertilization or by supplementary feeding. When the standing crop approaches the maximum, the population should be harvested and the whole process started again. In Israel and some

Figure 2.3 The River Garry, Scotland, formerly an important salmon river, after diversion of water for hydroelectric power.

tropical countries the rates of growth are extremely fast, and it is possible to obtain high annual production by taking two or more crops per year.

2.3.4 Industry

The availability of suitable water has determined the location of many large industrial areas in the world. The main factors concerned with such water are that it is available cheaply and is of appropriate quality and quantity. Water of high quality is required for certain industries and is often obtained from domestic water-supply systems. Some processes, however, can use water of poor quality and it can be cheaper to obtain this from boreholes, rivers, canals, private reservoirs, estuaries or even the sea, and use it untreated. Hopthrow (1963) estimated that in Great Britain about 35% of the public supply is used by industry but that much more than this is obtained from other sources. There are five major industrial uses for fresh water: processing, incorporation, boilers, cooling and firefighting. In addition, all industrial water users require domestic water for employee toileting, drinking and washing.

2.3.5 Hydroelectricity

The production of hydroelectricity involves a renewable resource, but can have major influences within the watershed concerned (Figure 2.3).

Figure 2.4 A major hydroelectricity dam in Norway.

Hydroelectric stations can be constructed only where the topography and water supply are suitable, but in mountainous areas with an adequate rainfall they make a significant contribution to power production. In 1989, for instance, though over 99% of the electricity generated in England and Wales was produced by thermal power, in Scotland more than 30% was generated by hydroelectric stations.

High rainfall, preferably with more in winter (when the demand for power is greatest) than in summer, sufficient altitude to give a head for generating purposes and the absence of prolonged freezing during winter are all factors favourable for hydroelectric production (Figure 2.4).

Various hydroelectric schemes have been developed according to regional differences in topography or power economics. In some large rivers, where adequate heads of water are available, base-load stations have been constructed with a capacity to suit low river flow. Other types involve storage reservoirs; the water available to these can be augmented by more brought in by aqueducts from neighbouring catchments. Such schemes may be single-stage, with one high-level reservoir, or multistage, with a high main reservoir above a stepped system of power stations, each with its own regulating pond. Pumped-storage schemes have become important in recent years for, with the need to maintain a constant load on nuclear power stations, their excess power, available at off-peak periods, can be used to pump water back into high storage reservoirs for hydroelectric stations.

Figure 2.5 Log rafts being floated down the Kemi River, Finland.

2.3.6 Navigation

Inland water systems have always been important for human transport from earliest times, especially where local topography or vegetation made overland transport difficult. Adequate navigation channels for large vessels coming from the sea to inland ports (and *vice versa*) are still important to industry and commerce in all parts of the world (Figure 2.5). Before the advent of modern air, rail and road systems, transport via inland waters was so important economically that there was substantial investment in canal waterways, mainly during the 18th and 19th centuries.

Unless they are exceptionally large (e.g. the Great Lakes of North America and the Rift Valley lakes of Africa) or can be connected by some form of canal system with other lakes or with the sea (e.g. the Caledonian Canal system in Scotland), natural lakes, though useful for local navigation, are rarely of importance in national terms. The main requirements of rivers for navigation are adequate supplies of water to maintain navigable channels and to counteract shoals in estuaries. If abstraction requirements of water undertakings and industry along the river are high, storage reservoirs may be necessary. Floods are a major navigation hazard in rivers, not only because of navigational difficulties but because the river bed is frequently scoured and deposited elsewhere in the form of shoals. Because of this and the small and shallow nature of many rivers during dry weather, large-scale navigation is restricted to the lower reaches of medium rivers, or to very large river systems (e.g. the Mississippi, Rhine and Danube) which have enormous natural channels and adequate dry-weather flows.

In many areas it has proved economic to extend the natural navigable systems, thus enabling ships to pass from sea ports right into inland industrial areas. This involves providing weirs, locks (for the passage of ships) and sluices to deepen the navigable channel and retain flood water; these barriers may also include turbines for hydroelectricity. This development and canalization of river systems has proved successful in many countries; for instance, in North America the St Lawrence Seaway, with only seven locks, developed a traffic of over 20 million tonnes per year within 3 years (Marsh, 1963). Such constructions may reduce the rate of flow of the river and increase deposition of sediment which entails the need for periodic dredging to maintain the navigable channel, with resultant disturbance to the substrate.

2.3.7 Recreation

In all parts of the world, the increasing human population, and the leisure time available to it, is placing an enormous demand on what is left of the natural countryside. This is especially true of fresh waters which are often the focus for a variety of recreational activities (e.g. sailing, power-boating, water-skiing, fishing, wildfowling, bathing and picnicking). It is unfortunate, but important to note, that for various reasons (including aesthetic ones) the most important waters for recreation are often not those already within urban areas, but those further away (Figure 2.6).

This is partly because of the natural preference for clean rivers and clear (usually oligotrophic) lakes as opposed to turbid, polluted and eutrophic waters. A study of the distribution of boatyards and building developments in England showed that the most popular areas for commerce and tourists lie in those river valleys where the greatest number and most important nature reserves are situated (Duffey, 1962).

Many wetland areas are being actively managed for the benefit of humans. Some are set aside and managed solely on the basis of their scientific interest, and recreation is actively discouraged. In other areas, various recreational activities are compatible with the aims of the reserve (Figure 2.7).

In the USA, natural wetlands are accepted as having a wide variety of uses including nature study, photography, hunting, fishing, snorkelling, boating, camping and picnicking, as well as economic pursuits such as lumbering, haymaking, grazing, mining, petroleum extraction and fur harvesting. While the primary interest in these areas may be wildlife or recreation, economic operations can yield substantial financial returns, and may even increase the value of the area for the wildlife or recreational activity concerned.

Figure 2.6 Boats for anglers at Loch Leane, Killarney, Ireland.

Figure 2.7 Canoeing to observe wildlife in Algonquin Park, Ontario, Canada.

Figure 2.8 Traditional willow basket traps for river lampreys, Finland.

2.4 THE PRODUCTS OF FRESH WATERS

2.4.1 Fisheries

Freshwater fish and to a lesser extent crustaceans and other animals are a major source of protein for some communities. Though published figures for the annual world catch from fresh waters are always substantially less than those from the sea, the figures rarely include subsistence fishing or sport fishing, which in some countries at least (e.g. the USA) exceeds the commercial catch from fresh waters. There is a remarkable similarity in the annual yield from marine continental shelf areas, which contribute about 80% of the total marine catch, and from fresh waters. The respective figures are 11 kg/ha and 12 kg/ha.

In some countries, the fish forming the basis of commercial fisheries have been introduced, thriving at the expense of, or in a niche unoccupied by, native species. Successful examples of such introductions are whitefish in Lake Sevan in Russia, and Pacific salmon in the North American Great Lakes. Most freshwater fisheries, however, rely on native species caught by methods that vary according to the species, water and tradition of local fisherpeople (Figure 2.8).

In large lakes, fish are caught by gill nets and traps, though in some waters seine nets or trawls may be used. In temperate areas, salmon, trout, charr, whitefish, pike, perch, pikeperch and other species are important commercially, while in tropical regions other species (such as cichlids, *Barbus* spp. and catfishes) are dominant. Most river fisheries rely on traps to capture both catadromous and anadromous species, though in some broad rivers and estuaries seine nets may be used. In temperate rivers, shad, salmon, trout, charr, and eels are the main species caught, while in tropical running waters *Vimba* and other lotic species are important.

2.4.2 Reed cutting

Reeds have been used since ancient times in construction and amongst the first manuscripts were those on paper made from *Cyperus papyrus*. Nowadays tall reeds, such as *Phragmites*, *Typha* and *Cyperus*, are used for thatching, fencing, windbreaks and the like, and more recently have been incorporated in the industrial production of insulating board, used in house building. In South America, the extensive use of *Scirpus totora* by the fishermen at Lake Titicaca, to build boats and floating islands on which to construct their reed huts, is well documented. The green *S. totora* is also harvested as fodder for their livestock (Levieil *et al.*, 1989).

2.4.3 Peat extraction

Hand cutting of peat to provide fuel for fires has been practised for centuries. The Norfolk Broads, in England, are secondary wetlands that have developed after flooding of extensive peat cuttings made during the middle ages. With good communications and easy access to higher quality fuels, hand cutting has greatly declined but in countries with large resources of peatlands, e.g. Russia, Finland and Ireland, peat is now extracted with heavy machinery, dried and used in powdered form to fuel power stations. In Ireland 80 000 ha is under exploitation (Goodwillie, personal communication). There is also small scale production of coke and activated charcoal.

Sedge peat is being exploited on an increasing scale for horticultural use, including private gardens, and small quantities are used in medicine. Peat is increasingly in demand in the chemical industry and is used as an absorbent material, for example in treating oil spills. In Russia it has been used to make insulating boards and no doubt alternative uses will grow, increasing pressure for peat extraction.

2.4.4 Other natural products and functions

Wetlands provide man with a variety of different plant and animal products that are useful as food. Northern peatlands furnish cranberries, blueberries and cloudberries, which, among others, are eaten by humans, birds and wild mammals. Edwards (1980) reports various algae and aquatic ferns being used as food in Asia and South America. Some of the many higher plants traditionally used as food are listed in Table 2.1.

Some plants, such as *Drosera* spp. and *Menyanthes trifoliata*, are used for medicinal purposes and further studies will probably reveal many more uses among the large number of species present. Wetlands are an important source of genetic material. Rice, which is the staple diet of about half the population of the world, has been extensively improved to meet local conditions, using wild strains found in wetlands. The harvesting of aquatic plants as fodder for domestic animals has already been mentioned. The trees and shrubs found in swamps and alluvial forests are used both for firewood and quality timber. On a world scale a wide variety of types of wood are present.

Emergent vegetation helps to stabilize shorelines, binding together the sediments and resisting the eroding effects of wave action.

Wetlands such as the Niger delta in Mali are obligatory spawning grounds for many species of fish, which move into them seasonally during periods of high water. They later act as nursery areas, enriched by the nutrients brought in by the river. When the waters recede the fish are caught in traps by the indigenous population and are an important part of the diet.

Large quantities of frogs' legs are now exported from some Asiatic countries (e.g. 300 tonnes from India annually: Löffler, 1990). In many Asian countries freshwater shrimps, crayfish and crabs are part of the human diet and the freshwater pearl mussel, *Margaritifera*, has traditionally been collected for pearls from rivers in the northern hemisphere.

Many waterfowl use wetlands permanently, to breed (Figure 2.9), to overwinter or both. Millions of birds overwinter in great wetlands like the Djoudj in Senegal, the Niger delta in Mali and Bharatpur in India but breed on the tundras of the northern hemisphere. Long before shooting waterfowl became a sport, ducks and other waterfowl were caught in nets or traps for food.

Wetland animals such as the coypu and capybara are still an important source of protein. Apart from their food value, certain animals, such as muskrat, coypu and beaver, have been hunted for their valuable furs, but the demand for these is now declining in developed countries. The large numbers of game animals such as antelopes on floodplains in Africa have become a well-known tourist attraction and in places such as the Kafue Flats in Zambia are being harvested to provide a sustainable yield as part of conservation management.

In South America, crocodile and alligator skins are still an important source of income to the local population and farming of crocodiles is reducing the culling of wild stocks in several tropical countries.

Table 2.1 Higher aquatic and wetland plants used as food by humans (adapted from Edwards, 1980)

Family	*Species*	*Parts used*	*Country*
Alismataceae	*Sagittaria trifolia*	Corms	China, Japan (also cultivated)
	S. sagittifolia	Corms	China (also cultivated)
Apiaceae	*Sium sisarum*	Roots	
Aponogetaceae	*Aponogeton* spp.	Tubers	
Araceae	*Colocasia esculenta*	Rhizomes	Egypt and tropics (mostly cultivated)
	Cyrtosperma chamissonis	Corms	South Pacific islands, SE Asia (also cultivated)
	Pistia stratiotes	Leaves	India
Brassicaceae	*Rorippa nasturtium-aquaticum*	Leaves	Widely used and cultivated
Convolvulaceae	*Ipomoea aquatica*	Leaves, stems	India, SE Asia, South China (also cultivated)
Cyperaceae	*Cyperus esculentus*	Tubers	Widely used
	Eleocharis dulcis	Tubers	Widely used (also cultivated)
Fabaceae	*Neptunia oleracea*	Leaves	Thailand (also cultivated)
Haloragaceae	*Myriophyllum aquaticum*	Tips of shoots	Java
Hydrocharitaceae	*Blyxa lancifolia*	Leaves	Thailand
	Ottelia alismoides	Entire plant without roots	Thailand
Lemnaceae	*Wolffia arriza*	Leaves	Thailand, Burma, Laos
Limnocharitaceae	*Limnocharis flava*	Leaves, stems, flower clusters	Malaysia, Java
Nelumbonaceae	*Nelumbo nucifera*	Whole plant	Widely cultivated in Asia
Nymphaeaceae	*Euryale ferox*	Fruits and seeds	India, China
	Nymphaea lotus	Stems, rhizomes, fruits, seeds	India, China
	Nymphaea stellata	Stems, flower stalks	India
	Victoria amazonica	Seeds	South America
	V. cruziana	Seeds	South America
Onagraceae	*Ludwigia ascendens*	Young shoots, leaves	Thailand
	L. repens	Young shoots, leaves	Thailand

Poaceae	*Hygroryza aristata*	Grain	North America
	Oryza sativa	Grain	North America
	Zizania aquatica	Grain	North America (introduced to tropics)
	Z. latifolia	Stems	Cultivated in China, Japan and Vietnam
	Phragmites australis	Rhizomes, buds	Indochina, Kirgistan
Podostemaceae	*Dicraeanthus* spp.	Stems, leaves	West Africa
Pontederiaceae	*Eichhornia crassipes*	Young leaves, petioles, flowers	Java
	Monochoria spp.	Mainly leaves	India
Potamogetonaceae	*Potamogeton* spp.	Rhizomes	India
Sphenocleaceae	*Sphenoclea zeylandica*	Young plants	Java
Trapaceae	*Trapa bicornis*	Fruit	Widely used and cultivated
	T. bispinosa	Fruit	Widely used and cultivated
	T. incisa	Fruit	Widely used and cultivated
	T. natans	Fruit	Widely used and cultivated
Typhaceae	*Typha angustifolia*	Rhizomes, flowers	Pakistan

Figure 2.9 Nesting pelicans in the Djoudj Marshes, Senegal.

2.5 OTHER ASPECTS

Wetlands have been used since time immemorial to provide grazing for domestic animals. In tropical regions they provide valuable nourishment during the dry season for cattle, sheep and goats, either by grazing the aquatic vegetation within permanent marshes or after the water has receded below the ground surface. The wetlands themselves benefit from the input of nutrients from the droppings of these animals. In some areas, where there still is abundant game, there may be competition with the domestic animals for this food resource, as well as for drinking water, which is the situation with the Masai in Kenya.

In the British Isles there has been widespread sheep farming in mountainous areas, where tracts of peat bog often provide summer grazing.

Traditionally, marshes where the vegetation is dominated by Graminae have been mowed to make hay. With the intensification of agriculture this practice is now declining in Europe and North America. This has considerable consequences for nature conservation, as a single mowing cut increases the diversity of plant species present.

In some tropical areas aquatic plants such as *Lemna, Salvinia* and *Pistia* are now being used in horticulture, either for composting or for mulching crops to conserve water.

2.5.1 Pollution reduction

Flood plains have an important natural function in taking up nutrients such as nitrogen and phosphorus from the flood water and incorporating them in the flood-plain plants, thus reducing problems of eutrophication downstream (Sanchaz-Perez *et al.*, 1991). Other elements, such as metals, may be retained by absorption on the substrate. The river waters may also deposit much of their sediment load on the flood plain, enriching its soils and reducing the problem of sediment build-up in any reservoirs downstream. In warmer countries, where biological activities are not limited by temperature, this property of purification has been exploited to apply year-round tertiary treatment to domestic sewage from small towns. Sewage that has received primary and secondary treatment is fed into a marsh or swamp by means of a perforated tube and water is drawn off from the opposite side of the wetland into the river system. Verhoeven and Van der Toorn (1990), referring to 22 studies carried out in Europe and North America, indicate high removal rates of 25–98% for phosphorus and 50–99% for nitrogen within different wetlands. Wetlands thus have a high future potential for purifying sewage from small towns and there is a considerable research interest in this. Clearly, sites of high conservation value should be excluded from this use.

2.5.2 Flood control

Wetlands play a significant role in flood control by acting as a 'sponge' that takes up excessive quantities of water resulting from heavy rainfall or rapid snow melt in the catchment, thereby reducing the degree of flooding and erosion downstream. This is an extremely important function in lowland areas and can avoid the need for expensive flood control structures. An added advantage is that the water held by the wetland is subsequently released gradually; thus excessive water level fluctuations, both high and low, are tempered downstream.

2.5.3 Research

Apart from their applied use to humans, wetlands and open waters, with their immensely varied conditions and diverse flora and fauna, have a high value for basic research on their intrinsic processes and properties. Peatlands are often very old (10 000–12 000 years), and store a wealth of undecomposed pollen grains, other plant material and insect remains, which permit the reconstruction of climatic and biological conditions throughout the period of their formation.

The diversity of freshwater resources, and their value to humans, emphasize the need to maintain them in good health for rational exploitation, for research and for future generations, as well as to safeguard those of high conservation interest.

3

Human impacts

3.1 THREATS TO FRESH WATERS

Only in recent years have humans become aware of the enormous damage being done to natural resources. Human influence on fresh waters is no exception. The conflict between the demand for large amounts of pure water on the one hand and the disastrous pollution of many waters on the other is only now forcing the issue with politicians. Enormous areas of wetlands have been drained for agriculture and forestry and changed out of all recognition.

For instance, in China about 30% of the lakes and 60% of the marshes have disappeared over the last 40 years as a result of conversion to agricultural land and the expansion of deserts. Because of the development of local industry and the increasing human population, about 80% of the existing wetlands along the Yangtze River are polluted and over 30% of the freshwater lakes in China have become eutrophic. More than 70% of all wetlands in China have been disturbed by human activities. A critical problem is the increasing desire to exploit wetland resources (fish, game, reeds, etc.), which offsets awareness of their natural functions.

Multipurpose river-basin projects seem the logical way to solve many problems. The following brief accounts indicate the major impacts of human usage on the ecology of freshwater systems.

3.1.1 Water demand

(a) Impoundments

Provided that there is an enlightened policy for river flow control, reservoirs can be beneficial within natural catchments. The result of building a dam across a valley is obvious locally, and the flora and fauna of the reservoir undergo rapid change from a lotic to a lentic community, which may take some years to stabilize. The effect of reservoirs is not only local, however,

Figure 3.1 Artificial, but of great value to local wildlife: a farm pond in New South Wales, Australia.

and changes occur in the river system below them. The outflow from a reservoir contains more plankton than the original stream, and stream animals feeding on them become commoner. On the other hand, the quantity of water in the outflowing river can be considerably reduced so that the total area is reduced (see Figure 2.3) and summer temperatures increase. Subsequently the amounts of plants and animals are reduced and the quality may change. The effects downstream can be far reaching: Drinkwater and Frank (1994) have shown that regulation and diversion of river water can have a deleterious effect on invertebrates and marine fish in coastal waters.

In other circumstances, lakes can exert a stabilizing influence on rivers (specifically so where water levels are controlled for flood prevention) and fluctuations in water level and temperature are reduced. In dry areas, artificial ponds of all kinds (Figure 3.1) can be an important source of water to local wildlife. In rivers with migratory fish, impoundments and weirs may affect ascent and descent; fish ladders and lifts are often built in an attempt to overcome this.

Dams of giant proportions (see Figure 2.1) have been built on some of the larger rivers of the world and, generally, the larger the reservoir (such as the Aswan High Dam) the greater the number of impacts on the environment – and the more catastrophic (here extending as far as the delta of the River

Nile), whereas very small reservoirs can be beneficial (Rzoska, 1976), reducing flood damage and playing the role of beaver dams. The river and, often, wetland sites that are drowned by a new reservoir are completely lost

In the Mississippi River valley of the United States, it is estimated that upriver dams have reduced the transport of sediments to the coast by half (Gagliano *et al.*, 1981). Dykes constructed along the lower river prevent flooding and distribution of sediments on former marshlands, which are being lost at rates as high as 100 km^2 per year.

(b) Hydroelectricity

All forms of abstraction of water have an impact on the hydrological functions of the catchment and, where the utilization of water is high, can deprive the natural aquatic habitats of water and thus be a threat to their value and even their existence.

Hydroelectric schemes have deleterious effects on local waters, because of abstraction, dams and turbines, water transfer and other activities. However, some fish are less affected than others by such schemes, and there is evidence that plankton-feeding fish may be favoured by the fluctuating water levels, which impoverish the main feeding grounds of benthic feeders in the littoral zone. Fluctuating water levels devastate the littoral flora and fauna (Smith *et al.*, 1987) and the benthic feeders, which occupy the littoral area, are adversely affected. As plankton are less affected, plankton-feeders still have an adequate food source.

(c) Abstraction

The effect of this varies in extent, except with total abstraction, where the results are obvious and disastrous. More often, only partial abstraction occurs, with variations in effect from year to year and place to place. In standing waters subject to rapid fluctuations in level caused by pumped-storage or flood-control projects, the shoreline experiences similar changes to those in abstracted rivers. There is a great reduction in macrophyte vegetation and in invertebrates that cannot withstand desiccation. Consequently, the shallow littoral areas of abstracted lakes and rivers, normally the richest zones, have poor production and specialized communities restricted to organisms that can withstand periodic desiccation (e.g. *Polygonum amphibium*) or are highly mobile and can keep pace with water level changes (e.g. *Gammarus lacustris*).

3.1.2 Land use

(a) Agriculture

The clearing of forests increases the runoff of surface water and the rate of soil erosion, with subsequent silting and nutrient increase in the waters draining such areas. Most types of cultivation lead to loss of soil and nutri-

Figure 3.2 Intensive land use in the lower valley of a river in Norway.

ents. Schogolev (1996) found that, over a period of 50 years in the valley of the River Dnestra, there was heavy soil erosion due to steep catchment slopes and the development of intensive agriculture in the vicinity of stream banks. Over this period, an estimated 13×10^6 m^3 of soil was deposited in the delta of the river, reaching depths there of 0.5–2.5 m.

Deficiency of nutrients is commonly overcome by the regular addition of agricultural fertilizers and these too tend to be washed off and affect the nutrient status and ecology of waters into which they drain. Rapid eutrophication caused by increased nutrient input from fertilizers and sewage effluents is one of the major problems in the management of fresh waters today. Peat bogs, being extremely deficient in nutrients, can be altered chemically if fertilizers or lime from aerial applications on farmland or forests are allowed to drift on to their surface, allowing atypical plant species to colonize.

The large quantities of rice grown in subtropical regions has meant that many of their wetlands have been modified into rice fields, to the detriment of their flora and associated fauna. The fauna has also suffered from the application of pesticides, as well as modification of the habitat, and is correspondingly reduced in diversity.

Large areas of wetland throughout the world have been drained to produce agricultural land (Figure 3.2).

In the Netherlands and Belgium nearly half the land surface was covered by bogs 2000 years ago. With the development of drainage techniques, large-scale drainage of the fens in the west of the Netherlands took place between the 12th and 14th centuries, and today only 36 km^2 remains intact in the Netherlands and about 18 km^2 in Belgium and Luxembourg.

In the 17th century, Dutch engineers were brought to Britain to apply their techniques to drain the extensive fenlands of East Anglia, and more recently the Vernier marshes at the mouth of the River Seine in France have been drained in the same way. Fournier and Wattier (1979) reckoned that 100 km^2 of wetland was being damaged or destroyed each year in France. This drainage, by means of ditches, canals and pumping, has extended to the present day and is very efficient. It completely transforms the wetlands into arable farmland and it is doubtful if they could ever be reconstituted. Less severe draining of the marshes and fens may result in a mosaic of grassland, arable ground, marshes and bogs and waterways, as in the west of France, and this does less damage to the environment, which may still contain a high natural history interest, though modified.

In north-west Europe, attempts have been made to drain blanket bog by ditching to provide grazing for sheep. This can result in the reduction of wetland plant species such as *Sphagnum* spp. and the increase of plants favoured by drier conditions such as *Molinia, Calluna* and *Nardus*. Where drainage succeeds, the ground may be ploughed, fertilized and reseeded to produce grassland with the help of agricultural subsidies, but this type of bog is difficult to drain because of its physical features and, with time, may revert towards its original state.

Grazing domestic stock can have a considerable effect on aquatic vegetation and indeed is used as a management tool (Chapter 6). In countries with annual dry seasons, as in Africa, cattle, sheep and goats are driven on to the drying-out wetlands and graze the emergent vegetation down to ground level. Only a few plant species, e.g. *Phragmites australis*, can withstand this over a number of years and the size and vigour of most plants is greatly reduced.

(b) Afforestation

All stages of forestry – from ground preparation and planting up to the mature crop and felling – have impacts on fresh waters (Maitland *et al.*, 1990). Depending on the stage, physical aspects affect:

- stream hydrology, by
  - increased water loss through interception by and evaporation from the forest
  - higher flood peaks and lower drought levels;
- sedimentation in streams and lakes by eroded material;
- reduced summer water temperatures from tree shading.

Chemical changes from afforestation include:

- increased nutrients from leaching and fertilizers;
- acidification from air pollutants leading to high aluminium levels associated with acid rain.

These effects combine to affect freshwater plants and animals. Changes in hydrology and water temperatures make conditions more extreme for biota. Turbidity decreases plant growth and increased nutrients increase algae. Acidification affects plants and invertebrates and may eliminate fish. Amphibians and birds may be reduced in number or eliminated.

Deciduous trees can contribute to the utilizable organic material by leaf fall into streams and rivers, but the needles of resinous trees may form sterile layers in forest streams.

With the development of deep ploughing techniques in the 1950s and the rising price of land, it became practical to plough peatlands for conversion to forestry. This has had a considerable but patchy impact on blanket bogs, often covering them with single-species stands of conifers of little or no biological interest. This is still one of the major threats to some of the finest blanket bogs in Scotland. In Finland, large areas of the peatlands have been drained for forestry, leading to a general increase in tree cover, loss of landscape heterogeneity and species diversity (Pakarinen, 1994).

Populus and *Eucalyptus* trees have been planted in marshy areas throughout the world to produce a rapidly growing crop of timber and at the same time to help marsh drainage by transpiration of water.

Eventual felling of forests in upland areas causes accelerated runoff of rainwater and degradation of formerly stable river beds. This can be catastrophic in mountainous areas of the tropics with high rainfall producing severe erosion of the river channels and flooding and loss of human life downstream. Brunig (1975) found that soil erosion may increase from a low level of 0.2 t/year/ha under virgin forest conditions to 600–1200 t/year/ha in areas clear-felled.

(c) River canalization

Straightening and canalization of river courses to prevent flooding has degraded many rivers in all parts of the world, as is pointed out in section 2.5.2. Accelerating water flow velocity and deepening erosion of the main channel leads to continued lowering of the water table, with detrimental effects on wetlands situated on the flood plain. These measures, combined with the construction of dykes (Figure 3.3) to prevent the river from discharging on to the flood plain, lead to the containment of water in the river channel during spates and prevent the natural dissipation of the kinetic energy contained in the fast-moving water mass (Pearson and Jones, 1975).

The problems created are thus transferred downstream. The flooding of neighbouring grasslands and fluvial forests is important both for releasing

Figure 3.3 Straightening the river course and dyking with boulders has destroyed the natural characteristics of the River Moselotte in France.

energy and removing silt from the river and for supplying fertile silt to the flood plain, but cannot now take place. The flood plains are also important for the purification of river waters through the uptake of nutrients and other chemicals.

In France, in both 1993 and 1994, the River Rhone, which has been progressively trained over many years, burst the banks protecting the Camargue for the first time since the dykes were built over 120 years ago. On the River Loire, in France and the River Ouse, in England, the rivers are dyked, but when the water rises to a certain level the excess water is allowed to spill, by special spillways, on to the original floodplain, thus reducing the chances of breaching of the dykes.

3.1.3 Pollution

Various definitions have been attached to the term pollution, but as it affects fresh waters it can be described as the discharge into a natural water of materials (usually waste products) that adversely affect the quality of plant and animal life there. The most important contribution comes from human activities in three fields: agriculture, industry and domestic waste disposal. The importance of the inter-relationship between fresh water and the disposal of human waste cannot be stressed too strongly. As human populations and activities on earth increase, more and more fresh water is required; most of this is polluted before being returned to the water course, and in many cases a major use of the water is to flush away wastes (Klein, 1957). As more water is used, so are more natural waters polluted; but with the demand for water, more and more polluted systems are being used for water supply, thus involving elaborate and expensive purification plants. An understanding of pollutants, their treatment and their effect on freshwater systems is an important part of the interpretation of freshwater ecology today.

(a) *Industrial and domestic sewage*

The influence of polluting substances on natural waters is variable according to local conditions and organisms within the water concerned. Pollutants can act in three main ways: by settling out on the substrate and smothering life there, by being acutely toxic and killing organisms directly, or by reducing the oxygen supply so much as to kill organisms indirectly. Since even clean cold water holds only about 12 mg/l oxygen, there is never a great deal available compared to air. Pollution may also be caused in other ways – by the addition of substances that may act as acids, alkalis or as nutrients. Radioactive substances, tainting of domestic water supplies and alteration of water temperatures are other examples.

Effluents with high suspended solids are typical of mining industries, poorly treated domestic sewage and various washing processes. Most of the solids settle out soon after discharge at a rate dependent on their size, density and local current conditions. The effect of inorganic particles is mainly a physical one, but plants and invertebrates may be completely covered and destroyed. Fish often die through their gills becoming clogged. If the particles involved are organic, their decay may add the problem of deoxygenation to that of alteration of the substrate.

The impact of toxic substances on organisms in natural waters is complicated by the fact that different species have varying resistance thresholds to poisons (which may act variably at different temperatures) and that some poisons are cumulative in their effect and others are not. Most toxic substances originate from industrial processes, though some arise from mining and agriculture.

Organic materials in sewage effluents are a source of major pollution of fresh waters. Though these effluents often contain plant nutrients, these cannot be utilized for some time because of the high oxygen demand of the decomposing organic material. In extreme cases, especially in lakes and slow-flowing rivers, so much oxygen is used up that anaerobic conditions result and no organisms other than bacteria and some fungi can exist. In less severe cases, species with low oxygen requirements (e.g. tubificid worms and red chironomid midge larvae) can exist and indeed, in the absence of predators and competitors and with abundant supplies of organic material for food, may build up dense populations. It is common, especially in running water situations, to find a sequence of changes from the area of greatest pollution near the effluent to cleaner water further away. With improving conditions as organic material is oxidized, there is a return to a natural flora and fauna, though the quality and productivity may be influenced by the nutrient salts present.

(b) Eutrophication

Through the passage of time, fresh waters naturally tend to become silted up and successional changes accelerate through the availability of potential nutrients locked up in deposits and released when shallowness permits their utilization, predominantly by macrophytes. This process has been greatly accelerated by eutrophication due to human activities. In lowland areas with fertile soils, intensive arable farming with heavy applications of fertilizers, particularly nitrates, leads to enrichment of ground water and runoff. Changes in the submerged vegetation of many of the Norfolk Broads in England in the 1950s, including their complete disappearance in some Broads, have been related to eutrophication and silting (Morgan, 1972). *Nuphar lutea* was the most resistant species and the last to disappear. Eutrophication occurs even in some water bodies in nutrient-poor areas, as a result of afforestation. This involves a considerable amount of ploughing and drainage and the application of chemical fertilizers to land newly prepared for tree planting, while the rehabilitation and reclamation of hill land for pasture also results in substantial amounts of fertilizer runoff into feeder streams. As well as agricultural fertilizers eutrophication also results from domestic sewage and other nutrient-rich effluents. Important nutrients are derived also from fish farming (from fish urine and faeces and from waste feed as it drops to the bottom). Principal among the nutrients concerned are compounds of nitrogen and phosphorus and there have now been many studies of the impact of these on the eutrophication of fresh waters.

Many of these nutrients quickly become bound up as part of the benthic biomass and the lake ecosystem in general. This is one of the principal features of the eutrophication process. Instead of a simple system, with a constant addition of nutrients and a constant dilution through the inflow–outflow system of the lake, there is actually considerable accumulation of nutrients within the ecosystem, both in its living components (plants, invertebrates and fish) and especially within the bottom deposits. Attempts to reverse the situation are often difficult because of the enormous store that has built up within the lake (Björk, 1972; Andersson *et al.*, 1975).

The effects of such eutrophication are various (Maitland, 1984; Henderson-Sellars and Markland, 1987) and include increased algal growths, potential deoxygenation of the lower cooler layer of water during stratification in summer and under ice in winter, and a tendency for the fish community to change from one dominated by salmonids to one where coarse fish predominate. Eutrophication is likely to be the main cause of the

Figure 3.4 Cockenzie Power Station, Scotland, one of many sources of atmospheric pollution and acidification.

extinction of fish in many lakes, e.g. the vendace, *Coregonus albula*, in the Castle and Mill Lochs in Scotland and the smelt, *Osmerus eperlanus*, in Rostherne Mere in England (Maitland and Lyle, 1991).

(c) Acidification

Acid deposition, arising mainly from the burning of fossil fuels (e.g. in coal and oil power stations, Figure 3.4) has resulted in severe damage to fish in Canada, the USA, Scotland, Norway, Sweden and other countries (Almer *et al.*, 1974; Maitland *et al.*, 1987).

Salmonid fish are particularly vulnerable (Figure 3.5), but most fish species have been affected and there are also major changes in the flora and fauna (Battarbee, 1984; Eriksson, 1984).

One of the characteristic features of acidification on fish is the failure of recruitment, manifested in an altered age structure and reduced population. This reduces intraspecific competition for food and temporarily results in increased growth or condition of survivors. As well as pH, the total ion content of the water is important to fish survival.

In France, on crystalline rocks, brown trout *Salmo trutta* have disappeared from streams where the pH has fallen below 5.6 and the aluminium concentration, released by the acidity, is > 200 ppb (Probst *et al.*, 1990). There is concern for the future of many systems in the poorly buffered areas of the

Figure 3.5 One of the results of acidification: deformed brown trout.

northern hemisphere if acidification continues. In addition, the invertebrate population is severely reduced by acidification, certain Plecoptera being the most resistant. In the most acid waters (some of which are naturally acid), all invertebrates and most plants are eliminated. The effects of acid rain on already acid bogs has yet to be established.

(d) Heated waters

With the high production of electricity from thermal power stations, the temperature regime of many natural waters has been significantly influenced by heated effluents. Relatively little is known about their influence on natural communities; it is likely that high temperatures will kill some stenothermic species and favour the development and reproduction of others. Also, the effect of organic effluents and some toxins is likely to be increased. In some temperate areas, species restricted naturally to tropical waters (e.g. the guppy *Poecilia reticulata*) have become established in the vicinity of heated effluents. The main effects of heated effluents as far as pollution is concerned are that warmer water holds less oxygen than cooler, and decomposition processes are speeded up.

3.1.4 Fisheries

(a) Commercial fisheries

The commercial harvesting of native freshwater fish need not have a harmful influence on the waters concerned. Indeed, because it is in their

Figure 3.6 A trout fisherman in Michigan, USA, one of the many millions of anglers worldwide.

interests to avoid contamination, fisherpeople act as a strong force against pollution and other influences. Occasional harm may be done to fish populations by overfishing, or where poisoning is carried out to collect some species or to control undesirable types. Efficient sustainable cropping probably has relatively little effect on the system as a whole.

(b) Sport fisheries

There has been increasing controversy in recent years concerning the impact of angling on aquatic wildlife (Figure 3.6).

A central problem concerns litter, which, in addition to being unsightly, has a serious impact on birds and mammals because of hooks and monofilament line in which they become entangled. Over the years, large quantities of lead, used as weights by anglers, have accumulated in lake and river sediments in European and North American countries. These are taken up, during gritting, by aquatic birds, causing saturnism, which can be lethal. In England, the population of mute swans declined markedly because of this. Sears (1988), in a study in the Thames valley and its environs in England, found that lead poisoning accounted for 94% of the deaths of local mute swans. Mudge (1983) suggested that 8000 mallard die each winter in the United Kingdom, while in the United States, Bellrose (1959) estimates that 1.4–2.6 million waterfowl die annually from this cause.

The presence of anglers often disturbs wildlife. Anglers can alter habitat, either unintentionally (e.g. by trampling down vegetation) or intentionally (e.g. weed cutting and bank clearance). Anglers may also impinge directly on aquatic communities by poisoning unwanted fish or shooting predatory birds and mammals. The transfer of alien fish species into new waters, either intentionally for sporting purposes or through the casual release of excess livebait, can have serious impacts on the native fish populations or on prey organisms, as with the release of pikeperch *Stizostedion lucioperca* and Danube catfish *Silurus glanis* in Europe.

(c) ***Aquaculture***

Aquaculture, which has been practised for centuries in south-east Asia, has greatly increased this century in many parts of the world, e.g. Africa (*Tilapia* spp.), Europe and North America (salmonid fishes). In many cases natural freshwater systems are modified by impoundments into which extraneous material is added in the way of fertilizer or fish food. Techniques have been developed to give very high rates of production, often combining aquaculture with agriculture. Ruddle (1980) reports two rice and eight fish harvests a year from fields in Java. Few of the original species of flora and fauna can resist the new conditions and the effluents from many fish farms, highly charged in organic matter, can cause severe pollution in rivers that receive them.

The recent increase in the number of fish farms around the world has posed many environmental problems. Solids from waste food and faeces pass into lakes or rivers, silting the bed and deoxygenating water. Nutrients which leach from fish feed, fish urine and faeces and from waste feed include nitrogen and phosphorus; these cause eutrophication and other problems. Chang (1994) claims that in China rotational culture of fish and macrophytes, in enclosures in natural lakes, actually reduces eutrophication, as the plants use excess nutrients accumulated in the sediments from the fish that preceded them.

Fish farms can be a source of disease to wild fish (Dolmen, 1987; Stenmark and Malmberg, 1987); fish farmers disclaim this but are themselves the first objectors to new fish farms near them. Many farmed fish find their way into local streams and interact with native fish by competing for space and food and predating eggs and young. Fish farms import various strains of fish from abroad and develop domestic races with characteristics unlikely to be advantageous in the wild. Fish escape or are introduced to the wild in such numbers that they may upset the genetic integrity of native stock or create problems of interspecific competition or direct damage to the habitat. American crayfish, which have escaped in many parts of western Europe, are causing damage by burrowing into the embankments of ponds and canals.

The main groups employed for aquaculture are fish, molluscs and crustaceans, which are often imported from other parts of the world.

(d) ***Dumping waste***

Small lakes and wetlands are often used as a convenient place to dump domestic and industrial rubbish (Figure 3.7). This not only disfigures and pollutes the environment but leads to eventual infilling and loss of habitats – particularly near large towns.

3.1.5 Recreation

Some recreational uses of fresh waters, such as bathing, have little effect other than in certain special cases, and difficulties arising are often due to conflict between the different types of recreation involved rather than their influence on the aquatic system. However, many recreational uses of fresh waters (wildfowling, angling, sailing, power-boating and water skiing) can cause pollution and disturbance to certain plant and animal species by actively killing them (wildfowl or fish), disturbing or frightening them away – an important problem with roosting and nesting birds. Mikola *et al.* (1994) investigated the effects of boat disturbance on broods of velvet scoter, *Melanitta fusca*, in Finland. Disturbance increased the distance between ducklings and the time used for feeding. The frequency of predation by gulls was 3.5 times higher in disturbed than in undisturbed areas. In southern Thailand, Pierce *et al.* (1993) found that aquatic birds reacted differently to different fishing methods, beating with seine netting causing the greatest disturbance. Yalden (1992) found that the presence of anglers around a reservoir in England reduced the breeding population of sandpipers *Acticis hypoleucos*. Areas of shoreline (Figure 3.8) and stands of macrophytic vegetation may be affected by trampling or the frequent passage of boats (Sukopp, 1971).

Accumulations of lead from boats and oil from their motors (Figure 3.9) are causing considerable pollution in many rivers and lakes and saturnism in birds from lead shot.

3.1.6 Climate change

There is increasing evidence that human activities of various kinds are altering the atmosphere to such an extent that global warming may create

Figure 3.7 Domestic waste dumped in a pond.

major climatic changes over the next few centuries. The most certain changes seem to be a rising sea level and a general rise in atmospheric temperatures, especially at high latitudes. Changes in precipitation, wind and water circulation patterns are also likely but their nature is uncertain. These changes are highly likely to affect aquatic habitats and organisms and a number of scenarios are possible (Maitland, 1991). Everywhere there is likely to be a shift of southern species to the north and a retreat northwards of northern species. In the open sea, changing temperature and circulation patterns are likely to affect pelagic, demersal and migratory species. Along the coast and in estuaries increased sea levels will create many changes to shallow water systems and produce problems for humans in low-lying areas. In wetlands and open waters, as well as the changes related to latitude, there are also likely to be parallel changes related to altitude, with cold-adapted species moving into higher cooler waters and their place being

Figure 3.8 Denudation of the shoreline of Loch Morlich, Scotland, caused by erosion of the plant cover by excessive recreational trampling.

Figure 3.9 Power-boating on the River Aar, a glacial river in Switzerland.

taken in the lowlands by warm-adapted species. In rich lakes in summer there will be an increasing tendency to hypolimnetic anaerobic conditions, with 'summer kill'; there will be less freezing in winter and so a lesser tendency to 'winter kill' in these lakes.

The progression northwards of warmer-water species will depend on their mobility and, for species without a marine or an aerial dispersal phase, on interconnections between water bodies. Restrictions in movement may lead to diminution in species diversity in certain areas, with mortality of the original cold-adapted species and limited colonization by warm adapted species. This may be particularly so for isolated peatland habitats.

3.1.7 Extraction of materials

Commercial peat digging from bogs (section 2.4.3) usually leads to their complete destruction and when extraction is terminated the land is usually put to either agriculture or forestry. Small-scale digging by hand, which has been going on for centuries, may produce small pools within a bog system which very slowly take on a natural aspect and produce habitat diversity, although the integrity of the system obviously suffers. Where large cuttings are flooded, shallow lakes may be formed, such as the Norfolk Broads and in the Netherlands, which produce a new freshwater habitat, but of an entirely different character.

Similarly, gravel workings on flood plains completely destroy any existing wetlands but usually result in flooded gravel pits, which, with management, can be of conservation value (section 6.3.4).

3.1.8 Burning

Burning of peatlands, which encourages new tender growth of the vegetation, has been used to improve rough grazing for sheep. Although this results in fresh palatable growth, burning reduces the vegetation diversity and burns that are too hot can affect the peat, producing a solidified surface not suitable for plant establishment (Rennie, 1957). *Sphagnum* spp. are the most susceptible to burning but lichens, such as *Cladonia florkeana* and *C. chlorophaea*, and an algal skin of *Zygognium*, are early colonizers of the burnt surface. *Erica tetralix* and *Narthecium* become abundant where fires are less severe, and *Scirpus caespitosa, Eriophorum vaginatum* and particularly *Molinia caerulea* develop on raised bogs (Goodwillie, 1980). Diversity of the plant community can be reduced by as much as 80%.

A physical impact of the combination of grazing and burning is the breakdown of the peat cover on mountain summits. Loose peat is eroded away causing gullying and leaving raised islands of peat stabilized by vegetation.

In marshlands where reed cutting is carried out to provide material for thatching, fencing and other purposes, the reed beds are often burned during the dormant season to produce clean early growths, with consequent decline in both floristic and structural diversity (Haslam, 1969). Late burns in the springtime can cause a great deal of damage to animals, particularly hibernating arthropods and nesting birds.

3.1.9 Control of disease

Pesticides to control vectors of disease, such as mosquitoes, black-flies and snails, have been used on a large scale in many wetland areas. It is rarely possible to do this in a species-specific manner, so that many other invertebrate species are killed at the same time.

The mosquito fish, *Gambusia affinis*, which was introduced into Europe and North Africa to control mosquito larvae, has proliferated in warmer areas. It appears to have exterminated the Mediterranean toothcarp, *Aphanius fasciatus*, over much of the latter's range and diminished the populations of related species around the Mediterranean.

3.1.10 Introduced species

Introduced aquatic plants have caused many problems in fresh waters throughout the world. Classic examples are *Eichhornia, Pistia* and *Salvinia*, natives of South America, which are now found throughout the tropics, shading out native plant species, deoxygenating the waters below their mats and blocking rivers, canals and the turbines of power stations. *Elodea canadensis* introduced into Europe has proved to be a very successful competitor with other plants and *Fallopia japonica* (previously known as *Reynoutria japonica* or *Polygonum cuspidatum*), *Impatiens glandulifera* and others, introduced as garden plants, are invading river banks throughout Europe, forming a dense monoculture that suppresses other plants and makes access to the river difficult (de Waal *et al.*, 1995).

One important aspect of biodiversity among fish communities is the role of introduced or translocated fish, of which 114 species have been introduced to Europe. This is a continuing process and many species, some of them potentially very harmful, are extending their ranges. A major point to remember here is that, whereas many other impacts can be almost completely reversed (e.g. pollution, channelization) to give a natural system, this is very rarely the case with introduced species once they have become established.

Most introductions are rarely needed or justifiable and many can be potentially damaging (Ben-Tuvia, 1981). The pikeperch was introduced to the Great Ouse Relief Channel in England in 1963 and subsequently spread to the Fenland Drain system, where changes in the native fish fauna are said to have been dramatic. In Lake Ymsen in Sweden, pikeperch (Figure 3.10) were introduced in 1911.

Figure 3.10 Pikeperch: an aggressive predator, the introduction of which has caused controversy in several countries.

The first harvest of this species was in 1914 and the yield rose until 1918 when it reached 13 tonnes; thereafter it dropped to 4 tonnes. As the numbers of pikeperch increased the numbers of bream, roach and perch declined dramatically. In Lake Erken, also in Sweden, introduction of pikeperch caused a decline in the catches of pike and perch; the removal of large numbers of pikeperch in commercial nets correlated with an increase in numbers of pike and perch.

In Great Britain, the ruffe, *Gymnocephalus cernuus*, is a small percid fish indigenous to the south-east of England, from where it has spread *via* canal systems to the English Midlands and eastern parts of Wales. The previous most northerly record appears to have been from the River Tees and the species was never recorded from Scotland until 1982, when ruffe appeared in Loch Lomond, 100 km north of their former area of distribution (Maitland *et al.*, 1983); it is now one of the commonest fish in the loch. It is abundant all over this large loch (71 km^2) and in its main inflow (the River Endrick) and outflow (the River Leven). It is believed that the ruffe was introduced to Loch Lomond by anglers from England, who frequently drive north to fish for pike, bringing various small fish with them to use as live bait. The impact of this new species on the existing community is uncertain, but unlikely to be beneficial, especially to the vulnerable powan whose eggs it eats in large numbers at spawning time.

Brown trout have been introduced all over the world from Europe as sport fish and have often been too competitive for native fish species which have either declined or disappeared. More recently, the introduction of the Nile perch, *Lates niloticus*, into Lake Victoria to improve the commercial fishery (Barel *et al.*, 1985) has had a more stunning effect as, possibly in association with increased eutrophication caused by agricultural runoff, this predator has eliminated most of the endemic species of haplochromine fishes from the lake. In terms of loss of genetic material, this is considered to be the biggest ecological catastrophe of the century. Over 200 species are believed to have been eradicated – the largest extinction of vertebrates this century (Goldschmidt *et al.*, 1993).

The mosquito fish *Gambusia affinis*, native to North America, has been introduced to many parts of the world to control malaria-transmitting mosquitoes. It has rarely been effective in controlling the mosquitoes but has proved to be a very successful colonizer and competitor with other fish species and in certain cases is thought to have eliminated them, as, for example, with *Aphanius fasciatus* in the south of France.

The painted terrapin *Chrysemys picta*, another North American species, has been imported in thousands into Europe for the pet trade. Unfortunately, when owners become tired of them, the terrapins are often released into local lakes and ponds. They are now established in many parts of France, in spite of efforts to prevent this practice and to control importation. The ecological implications are as yet unknown, but once more it is too late to reverse the process.

The North American ruddy duck, *Oxyura jamaicensis*, has now established itself in Western Europe, *via* escapes from captivity, and it is feared that it will hybridize with the rare whiteheaded duck *Oxyura leucocephala*, which is in danger of extinction. Coypu, *Myocastor coypus*, and muskrat, *Ondatra zibethicus*, introduced for their fur, have damaged vegetation and caused leaks in dykes by their burrowing. In the Norfolk Broads in England, coypu brought about an enormous decline in *Phragmites* and *Typha* spp. (Ellis, 1963). American mink, *Mustela vison*, have escaped from fur farms and established themselves in the wild. They are now preying extensively on all kinds of ground-nesting birds and are a major ecological threat in many areas.

Dangers also exist from the introduction of parasites or diseases, along with or separately from the host species. A good example is the nematode *Anguillicola crassus*, which attacks the air bladder of eels, which was introduced into Europe in 1982 and has spread rapidly – having a serious effect on the eel population, in contrast to the situation in its native Japan, where the eels are adapted to it.

The major impacts of humans on freshwater habitats and their consequences are summarized in Table 3.1.

Table 3.1 A summary of the main pressures facing freshwater habitats worldwide

Danger	*Effect*
Industrial and domestic effluents	Pollution; poisoning; blocking of migration routes
Acid deposition	Acidification; release of toxic metals
Land use (farming and forestry)	Eutrophication; acidification; sedimentation
Industrial development	Sedimentation; obstructions (including roads)
Warm water discharge	Deoxygenation; temperature gradients
River obstructions (dams)	Blocking of migration routes; sedimentation of spawning beds
Infilling, drainage and canalization	Loss of habitat, shelter and food supply
Water abstraction	Loss of habitat and spawning grounds; transfer of species
Fluctuating water levels (reservoirs)	Loss of habitat, spawning and food supply
Fish farming	Eutrophication; introductions; diseases; genetic changes
Angling and fishery management	Elimination by piscicides, diseases, introductions
Commercial fishing	Overfishing; genetic changes
Introduction of new species	Elimination of native species; diseases; parasites
Global warming	Loss of some southern or low-altitude populations; movement of species towards poles
Water transfer, canals	Transfer of species and disease
Canalization and dyking	Habitat loss; flooding
Extraction of peat	Destruction of habitat
Use of pesticides	Loss of biodiversity
Water recreation	Disturbance; habitat loss

4

Evaluation of conservation interest

4.1 INTRODUCTION

Aquatic habitats are susceptible to damage or destruction by so many physical and chemical processes that full protection can only be achieved when there is complete control of the catchment. This is rarely the case, except with small water bodies or sites within large national parks. Even then aerial pollution from afar, particularly from industry, can contribute to nutrient enrichment and acidification. Numerous attempts have been made, particularly in Sweden (Anon., 1991), to reverse the process of acidification, mainly by adding calcium carbonate, but although this increases the pH and alkalinity the studies have not been detailed enough to show whether the flora and fauna return to their original state. Thus, although this technique certainly alleviates the situation, we cannot yet say whether it can be used to protect the whole aquatic community.

In general the main means of protection of sites has been by the establishment of nature reserves or national parks but it is seldom that the whole catchment can be included, as this would usually increase the area of the reserve by many times. Thus the majority of protected sites are still vulnerable to activities outside their perimeter. This situation can only be alleviated by effective integrated land-use planning on a regional scale, taking into account the protected sites. This occurs in very few countries. However, in the wider countryside, the use of impact assessment techniques has been gaining ground in many countries as a valuable means of gaining information about particular habitats and assessing the impact of particular developments on them (Smith *et al.*, 1983).

Where a large freshwater body lies across the boundaries of several countries control becomes more difficult. An example is Lake Tanganyika, which has four riparian countries, the official language of two being French and that of the other two being English. There is still much to be done, and many problems to be overcome, to achieve adequate protection of this valuable site.

4.2 SITE SELECTION

In the past, many sites have been chosen for protection on an *ad hoc* basis because they were known to specialists and naturalists for their exceptional interest: limnological, botanical, ornithological, etc. This process was perhaps satisfactory for certain sites but it did not assess them against similar sites elsewhere nor did it position them within the range of ecological variation of that habitat. To do this requires a systematic approach of site survey, classification and evaluation that aims at highlighting the best examples for protection.

4.2.1 Site survey

Such a procedure is described in Ratcliffe (1977) for a number of different habitats in Great Britain. Teams of biologists set out to assess their particular habitat on a countrywide scale. For open waters and wetlands, field survey data were combined with information obtained from the national conservation agencies, limnological laboratories and universities. For open waters, the information consisted mainly of physical, chemical, botanical and zoological data (Morgan and Britton, 1977), and for wetlands, physical, botanical and ornithological data (Goode, 1977). Field surveys were carried out at potentially interesting sites for which there were insufficient data to determine their conservation value. Such sites were selected over a broad geographical and geological range, factors likely to affect their flora and fauna. Field samples were taken of the water, macrophytes, zooplankton and macroinvertebrates for later analysis. Information on fish and aquatic birds was usually available from angling clubs and ornithological societies. Further information was gathered in the field on such parameters as the dimensions of the site, substrate, water use, surrounding land-use and so on. To speed the process, these data were entered directly on cards (Morgan, 1990), as illustrated in Figure 4.1, which shows the field card for standing waters.

For example the substrate is divided into nine types and the extent of each is estimated as a percentage of the whole. Such field cards are invaluable *aide mémoires* for the collection of information in a systematic manner. The collection of the information in a numerical form facilitates analysis, particularly if computer analysis is to be carried out.

Once the analyses are completed and the data are available for every site, a comparison of sites can be made to determine their relative conservation value. It is important to note that comparison can only be made between sites of a similar type and not between habitats that are essentially different, such as lakes and peat bogs. It is therefore necessary to classify the sites into groupings with substantially similar characteristics, e.g. oligotrophic lakes, eutrophic lakes, ombrotrophic peat bogs and so on. The exact form of the classification may vary from one region to another. Some examples of such classifications are given in section 7.2 and Figures 1.6 and 4.2.

4.2.2 Classification of habitats

Two approaches have been made to the classification of aquatic sites for conservation and planning purposes. Traditionally the common procedure has been to construct a hierarchical classification, based on the author's experience and existing literature, and then to carry out field survey in order to allocate sites to groupings within the classification. More recently, with the arrival of computer science, the converse has been the case and extensive surveys have been carried out to provide the data with which to build the classification.

Cowardin *et al.* (1977) drew up a detailed classification of wetlands and deepwater sites for the whole of the USA, based on the literature and an inventory of wetlands carried out by Shaw and Fredine (1971). This is one of the most extensive classifications produced so far. It is designed for use with varying objectives and by people with different competences, and can accordingly be used at the appropriate hierarchical level. The further one descends in the hierarchy the more specialist is the information needed and the finer the differentiation between groups. At the most refined level of use, detailed field survey is required to obtain the necessary information. An important aim was to obtain uniformity of the definition of different types of aquatic habitat, so that where comparisons were made it was with reasonable assurance that they referred to similar types throughout the USA.

In remote areas of underdeveloped countries it may only be possible to make one visit to each site to obtain the data necessary for evaluation, as above. The time of year is important in relation to seasonal variations. Morgan and Boy (1982) used a more sophisticated approach to classify North African wetlands based on rapid field visits carried out during the wet season plus analysis of LANDSAT satellite imagery to show water fluctuations in the seasonal wetlands. The findings were divided into four data sets: physical, macrophytes, invertebrates and winter counts of waterfowl. Cluster analyses were made of each of the four sets in order to construct dendrograms. These were combined using a cross-matrix of resemblance to produce a classification based on the entire set of measured characteristics (Figure 4.2).

Correspondence analysis was applied to show the relationships of groups of sites to physical, chemical and biological components. In this way the classification is explained and the characteristic components necessary for objective assessment of the conservation value are defined. This is particularly useful in determining what is representative for a site. In this exercise a total of 11 distinct groups of wetlands were distinguished and

OPEN WATER HABITATS: STANDING WATER	SITE NAME: Loch of Kirkigarth 1 inch 7th series O.S. map No. 2

1 Abbreviated Site Name	1 · 9 · 20	2 RECORDER	RHB	3 DATE	28/10/7[illegible]
4 County	Shetland	21 · 5 Grid Ref.	23 N 4 1 2 3 7 4 9 6	6 100m sq. no. · 31	7 Area ha · 33 · 7·0

	Col.	Entry	A	B	0	1	2	3	4	5	6	7	8	9
8 Altitude	38	5 m		-50	-100	-200	-300	-400	-500	-600	-700	-800	-900	> 900
9 Perimeter	39	m km		10m	-50	-100	-500	500-1km	1-2km	-5	-10	-20	-50km	-100
10 Perimeter/Perimeter of circle of equal area	40		1-1.2	1.2-1.5	1.5-2	2-3	3-4	4-5	5-6	6-7	7-8	8-9	9-10	> 10
11 Maximum Length		600m					12 Direction		N.W - SE.					
13 Status	41		NNR	PNNR	WFR	PWFR	SSSI	Other Reserve	NP	AONB	Un-sch'd'l'd	Informal RA	RSPB Reserve	
14 Ownership	42		FC	NT	Common	Private	Public	NC	15 Water Authority			River Authority	River pur'f'c't'n board	County Council
16 Catchment Area	43	1·4 km² (ha)		10ha	-50	-100	1-2km2	-5	-10	-20	-50	-100	-200	> 200
17 Outstanding Interest	44			Aquatic	Botanical	Inverts.	Fish	Reptiles Amphibia	Mam'ls	Birds: Summer	Birds: Winter	Geological	Terrestrial	Other
			Expand if freshwater interest:											
18 Exposure	45	Moderate					N	NE	E	SE	S	SW	W	NW

19 Subjects on which information is available	Bathymetric Survey	Chemistry	Hydrology	Algae	Macrophytes	Zooplankton
	Benthic I'vrt'brts	Fish	Aquatic Birds	Publications	Photographs	Other

20 Sources of Information

21 Indicator Species

46	
	A
	B
	0
	1
Macrophytes	
Ceratophyllum demersum	2
Chara aspera	3
C. papillosa	4
Potamogeton lucens	5
P. praelongus	6
P. filiformis	7
P. pectinatus	8
Ranunculus circinatus	9

47	
	A
Macrophytes	
Sphagnum subsecundum	B
S. cuspidatum	0
Isoetes lacustris	1
I. echinospora ?	2
Nitella spp.	3
Subularia aquatica	4
Eleogiton fluitans	5
Lobelia dortmanna	6
Potamogeton polygonifolius	7
Ranunculus flammula	8
Sparganium angustifolium	9

48	
	A
Cladocera	
Holopedium gibberum	B
Hirudinea	
Hemiclepsis marginata	0
Glossiphonia heteroclita	1
Erpobdella octoculata	2
Gastropoda	
Amnicola confusa	3
Hydrobia ventrosa	4
Bithynia leachii	5
B. tentaculata	6
Limnaea auricularia	7
L. stagnalis	8
Planorbis carinatus	9

49	
	A
Gastropoda	
Planorbis corneus	B
P. complanatus	0
P. planorbis	1
Segmentina nitida	2
Viviparus fasciatus	3
V. viviparus	4
Bivalvia	
Anodonta anatina	5
A. cygnea	6
Pisidium amnicum	7
P. henslowanum	8
P. conventus	9

50	
	A
	B
Plecoptera	
Dinocras cephalotes	0
Perla bipunctata	1
Ephemeroptera	
Leptophlebia vespertina	2
L. marginata	3
Ameletus inopinatus	4
Tricladida	
Dugesia lugubris	5
Phagocata vitta	6
Dendrocoelum lacteum	7
Malacostraca	
Gammarus duebeni	8
Asellus spp.	9

51	
Corixidae	
Corixa scotti	A
C. castanea	B
C. semistriata	0
C. sahlbergi	1
C. punctata	2
C. lateralis	3
Fish	
Common Bream	4
Common Carp	5
Tench	6
Charr	7
Salmon	8
Sea Trout	9

52	
	A
	B
Birds	
Grey Lag Goose	0
Shoveller	1
Pochard	2
Coot	3
Great Crested Grebe	4
Pintail	5
Wigeon	6
Scoter	7
Red Throated Diver	8
Black Throated Diver	9

Figure 4.1 A field card for gathering information on temperate standing waters.

No.	Item	Col.	Unit	A	B	0	1	2	3	4	5	6	7	8	9
22	Mean Depth	53	m				< 1	1–2	2–3 ? (crossed)	3–5	5–10	–20	–50	–100	> 100
23	Maximum Depth	54	m				< 1	1–2	2–3	3–5	5–10	–20	–50	–100	> 100
24	Islands	55		Number:			Present	Absent (crossed)	Temporary	Permanent					
25	Ice Cover	56		?			Seldom	1 month	–2 mths	–3 mths	None				
26	Water Use	57		Research Education	Water Abstraction	Motor	Boating: Other pow.	Boating: Angling (crossed)	Boating: Shooting	Shore use: Angling (crossed)	Shore use: Shooting	Other	None	Future develop'mts: Deleterious	Future develop'mts: Unimportant
27	Land use of catchment	58		1 Upland Grassland (crossed)	90	2 Dwarf Shrub Heath		3 Acid Peat Bog		4 Conifer Woodland		5 Mixed or Broadleaf wood		6 Other Grassland	
				7 Arable (crossed)	10	8 Urban Industrial		9 Fen							
28	Land use of bordering land	59		1 Upland Grassland (crossed)	100	2 Dwarf Shrub Heath		3 Acid Peat Bog		4 Conifer Woodland		5 Mixed or Broadleaf wood		6 Other Grassland	
				7 Arable		8 Urban Industrial		9 Fen							
29	Outflow	60					Present (crossed)	Absent	Permanent	Intermittent					
30	Inflow	61				Ditches	Streams Rivers (crossed)	Flushes	Submerged Springs	Ground Water	Precipitation	Intermittent	Permanent		
31	Geology	62	[handwritten: O.R.S]			Rocks: Calcareous (crossed)	Rocks: Acid	Rocks: Other	Drift: Till	Drift: Sand, gravel	Alluvium	Peat	U'lying Rocks much ex.	Permeable	Impermeable
32	Origin	63				Natur'lly dammed Lake	Artific. dammed	Artific. excavated	Coastal Processes	River Processes	Pingo Kettle Hole	Corrie, Cwm, or Cirque	Glacial Trough (crossed)	Subsidence	Other
33	Shore Substrate	64		A Trees and Shrubs		B Emergent Reed Swamp		0 Bedrock	✓	1 Boulders	✓	2 Stones (crossed)	10%	3 Gravel	
				4 Sand		5 Silt		6 Organic Mud		7 Peat		8 Artificial Embankment		9 Other	
34	Aquatic Vegetation	65		1 Emergent		2 Floating Attached		3 Free Floating		4 Submerged (crossed)					
35	Trophic Status	66	[handwritten: poss. mesotrophic]				Marl	Base Rich	Base Poor (crossed)	Peat Stained	Brackish				
36	Pollution	67				None	Domestic Organic	Domestic Other	Indust. Organic	Indust. Toxins	Indust. Solids	Indust. Temp.	Agric. Organic	Agric. Fert. (crossed) [handwritten: ? little]	Agric. Pestic.
37	pH	68		4.0–4.4	4.5–4.9	5.0–5.4	5.5–5.9	6.0–6.4	6.5–6.9	7.0–7.4	7.5–7.9	8.0–8.4	8.5–8.9	9.0–9.4	> 9.5
38	Alkalinity (ppm $CaCO_3$)	69		0–1	1–5	5–10	10–20	20–30	30–50	50–70	–100	–150	–200	–300	> 300
39	Water Fluctuation	70					< 30cm (crossed)	–100 cm	1–2 m	2–5 m	> 5 m	Natural	Artificial		
40	Water colour	71				Suspended Matter (crossed)	Clear	Light Green	Dark Green	Pale Brown (crossed)	Dark Brown	Pale Yellow	Deep Yellow	Orange Brown	
41	Abundant Groups	72					Flatworms	Leeches	Asellus	Stoneflies	Snails	Gammarus	Tubifex		
42	Neighbouring Water bodies														
43	Annual rainfall	73	mm			< 600	–750	–1000	–1250 (crossed)	–1500	–1750	2000	2250	–2500	> 2500

44 General remarks (Conservation value etc.)

[handwritten: One of richest Shetland lochs botanically. ? P. rutilus recorded.]

Figure 4.1 Continued.

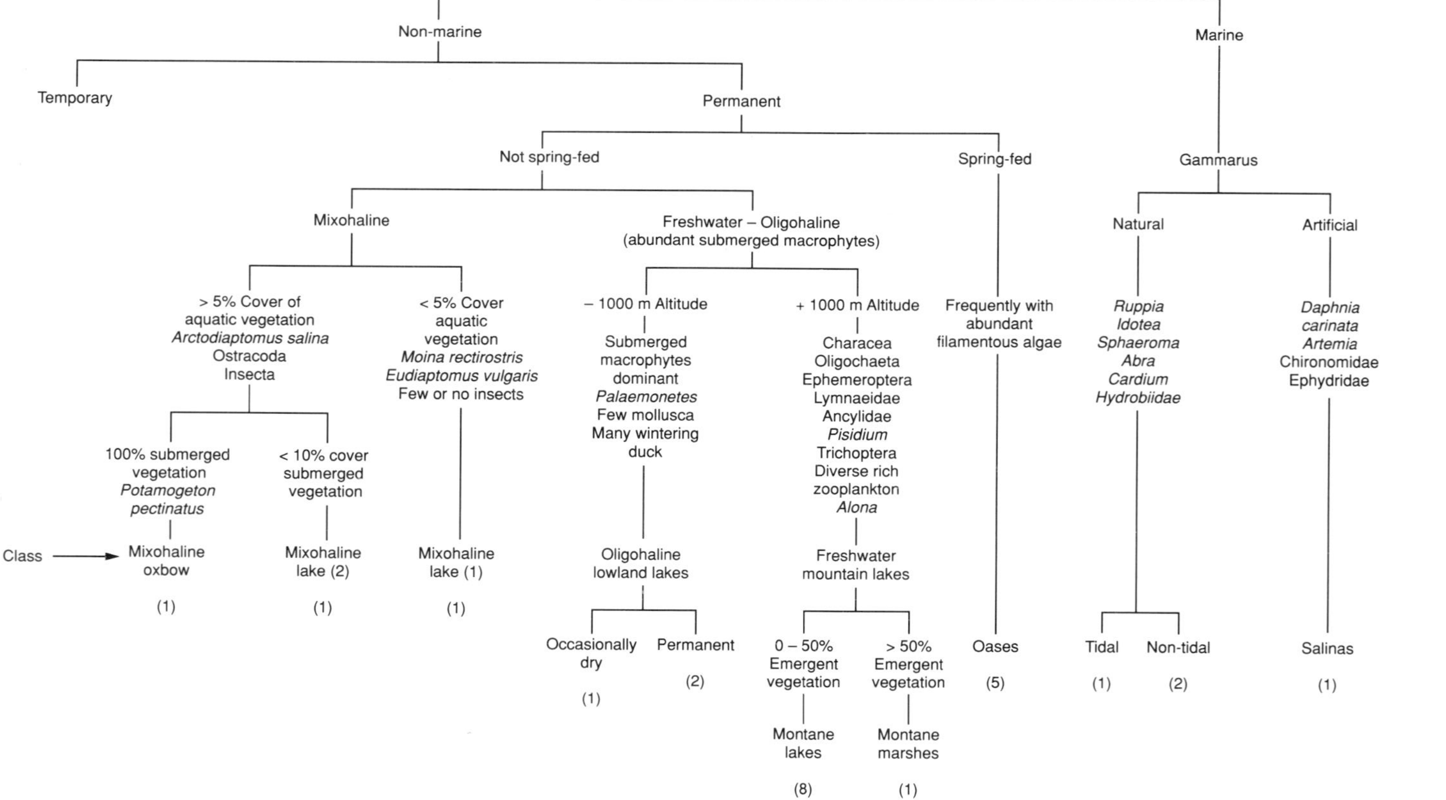

Figure 4.2 A natural classification of permanent waters in north west Africa, using the technique of cluster analysis (after Morgan and Boy, 1982).

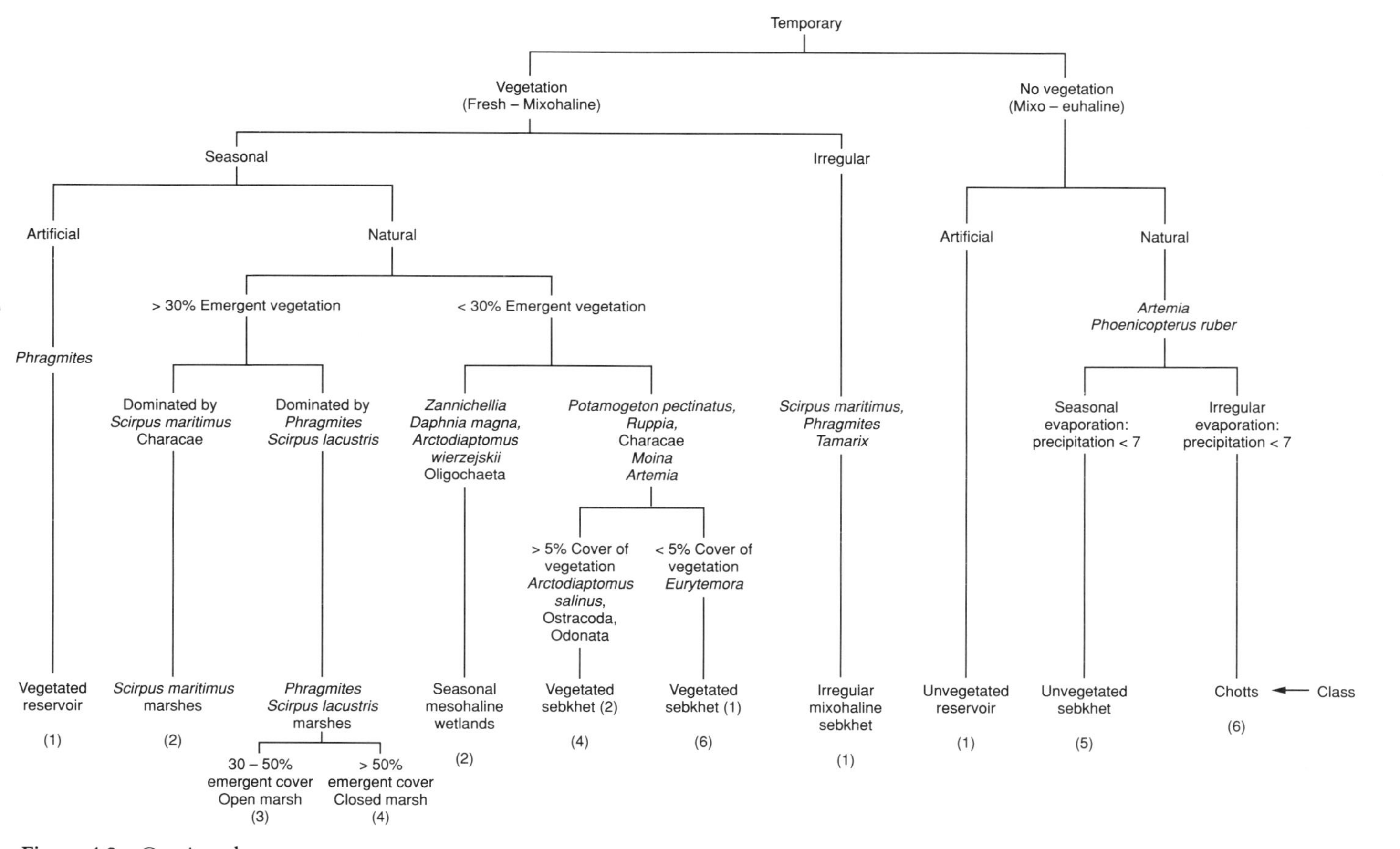

Figure 4.2 Continued.

nine individual sites that were unique within the survey (Morgan and Boy, 1982; Morgan, 1982).

Where data are already available for a large number of sites, mathematical analysis can be applied straight away. It is rare, however, that sufficient comparable data are available to do this and it is generally necessary to carry out supplementary field work. The rapid survey technique lends itself best to regional surveys in areas where little information is available, and where data must be collected quickly and in a uniform manner.

There exists a need for a detailed classification on a world scale to enable valid comparisons to be made of the international value of sites using agreed criteria. Outstanding areas, such as Lakes Baikal and Tanganyika, are obvious, but quality sites that are not so evident are less easily ranked. The terms local, regional, national and international importance are too easily applied without enough thought or guidelines. Advanced photo-interpretive techniques, involving aerial photographs and satellite imagery linked to computer analysis, allow rapid collection of certain parameters on aquatic sites. This permits statistical analysis of large quantities of information, as in the widescale survey of the wetlands of the USA. It is important to note that, again, comparison can only justifiably be made between sites of a similar type, as defined by the classification, and not between habitats that are essentially different. Recently there has been much discussion, in relation to limited funding for research and conservation, on methods of determining the minimum number and area of such sites which it is necessary to protect, for a particular area, in order to conserve the maximum representative diversity (Margules *et al.*, 1988; Saetersdal *et al.*, 1993).

4.2.3 Criteria for evaluation

Before a decision can be made on whether a site should be protected, or what level of protection is required, it is necessary to make an evaluation of the conservation value of the site. The premier nature reserves were chosen by 'instinct', as their exceptional values were obvious, but for less exceptional sites some rational form of selection is needed. Thus, various criteria have been applied by different authors. Those proposed by Ratcliffe (1977) for an assessment of the major habitats throughout Great Britain for the nature conservation review were found adequate for open waters by Morgan and Britton (1977) and for peatlands by Goode (1977) as part of this review. They have been widely used elsewhere and were recently adopted by the European Commission, with slight modifications. Widely used criteria, based on those proposed by Ratcliffe (1977), which can be applied to evaluate freshwater habitats, are:

- **Principal criteria**
  - Naturalness
  - Diversity of habitat

Figure 4.3 The River Racaca, New Zealand, a fine example of a natural river unmodified by humans.

- Species diversity
- Extent
- Rarity
- Fragility (viability)
- Representativeness

- **Supplementary criteria**
  - Recorded history
  - Site of important research (or potentially so)
  - Potential for educational activities
  - Contiguity with other protected sites
  - Species at the extremes of their geographical distribution
  - Feasibility of effective protection.

(a) Naturalness

This criterion relates to the closeness of the habitat to its natural state before being influenced by the activities of humans (i.e. intactness). Few fresh waters are entirely untouched by man. Those that are, are generally situated where the population density is low, such as high mountain lakes and rivers, or in countries like Iceland and the South Island of New Zealand (Figure 4.3).

Severe climates and the presence of biting insects such as midges (Ceratopogonidae) and mosquitoes (Culicidae) have saved many peatlands from exploitation. Throughout the world many rivers have been dammed for water supply and hydroelectric power, reducing the discharge downstream and regulating lakes so that they are subjected to unnatural water regimes.

Where there are large-scale fluctuations in water level in reservoirs, the flora and fauna of the littoral and sublittoral can be severely depleted (Smith *et al.*, 1987). In developed areas, rivers are being progressively straightened and embanked by engineers to protect developments and prevent flooding. The engineering works on lowland rivers, which prevent them from changing their course and spilling over into their flood plains, seriously restrict their natural function and may eventually lead to catastrophes. This has occurred recently (1993) on the Mississippi River in the USA, and the River Tay in Scotland, when exceptional rainfall caused both rivers to break newly engineered flood banks and flood adjacent townships. At the same time the structural variation of such rivers is reduced so that they cannot braid and differentiation into runs and pools is considerably reduced. In Ireland there has been such extensive exploitation of peat bogs for fuel that it is difficult to find a natural site of any size. Thus such a site can easily become of national value for that reason alone.

Pollution is perhaps the most general impact by man on former fresh waters, whether by point source industrial or domestic pollution, or by more diffuse agricultural input. Given the will and the resources, the former can be stopped in many cases but the chemical changes engendered in the ground water by the latter may take decades to reduce, with consequent difficulties for conservation.

Intactness of the catchment as well as that of the water body itself plays an important role. Oligotrophic sites are more susceptible to slight pollution or enrichment than eutrophic sites and consequently the intactness required in the catchment of the former type is much greater. Most oligotrophic waters are in relatively unpopulated uplands, but aerial applications of fertilizers (e.g. to forested areas) may pose a threat, as for peatlands, which are also poor in nutrients.

Motor-powered boats, whether for sport or tourism, can destroy vegetation directly through propeller damage (with loss of the accompanying fauna) and indirectly by creating turbidity, which reduces light penetration.

Artificial waters cannot be considered natural but with time may exhibit the fundamental characteristics of natural waters with a rich flora and fauna. Parts of the Norfolk Broads in England, a large complex of medieval peat diggings, are now considered to be of international conservation importance (Morgan and Britton, 1977) even though they have suffered from problems of eutrophication from agricultural nitrates. Flooded gravel pits become refuges for migrating or overwintering waterfowl and aquatic plants. These sites are often manipulated by man, both physically and by planting aquatic plants to produce a favourable habitat (Harrison, 1982). Similarly, some rivers that give a very natural appearance, such as the chalk streams of southern England, have been intensively managed for brown trout fishing by frequent weed cutting and dredging and removal of 'coarse' fish. However, they represent a diverse habitat no longer found in a natural state in Britain.

These examples show that a criterion such as naturalness cannot be used in a categorical manner but should be weighed in relation to other factors. A truly natural site should, however, have high priority.

(b) Diversity of habitat

Biological diversity (biodiversity) relates to the wealth of different life forms – microorganisms, plants and animals – that exist and the complex ecosystems they make up. Within each species there may be races and within races differences exist between individuals (genetic diversity). Species coexist to form communities, which combine as ecosystems that exhibit both species and habitat diversity. The extent of genetic diversity cannot usually be determined but estimates of species and habitat diversity are practical.

The latter has been widely used in assessing sites worthy of protection as it indicates the potential of the site and is fairly rapidly determined. In fresh waters, chemistry and physical conditions have a strong controlling influence on the biota (sections 1.3.1 and 1.3.2). Thus, habitat diversity greatly influences both the biomass and species composition. In general, nutrient-rich sites have the richest flora and fauna and nutrient-poor the poorest. The conditions in a river change where tributaries, running off chemically different geological formations, enter it.

A river with a diversity of physical features such as islands, oxbows, backwaters, rapids and gorges would take precedence over a uniform river with similar water chemistry. A lake exhibiting a variety of shorelines, from sheltered bays with fine sediments through a range of increasing particle size to exposed rocky shores, is likewise of potential conservation value. A range of depths, the presence of islands (of value to aquatic birds) and deltas are important. The latter in itself can show a considerable diversity when a well-developed seral succession is present, ranging from submerged plant communities to carr woodland. Similarly, a raised bog with a well-developed pool and hummock system merging into blanket bog or wet woodland is of high diversity value.

(c) Species diversity

Plants flourish in the warm and wet conditions of the tropics and the largest number of species are therefore found there. Generally, animal diversity is related to plant diversity. Tropical wetlands are therefore among the richest ecosystems in the world, whereas arctic and high-altitude habitats are species-poor. Certain taxa (e.g. bacteria and some invertebrates) are often not well enough known for a particular site, but better known groups, such as higher plants, fish, birds and mammals reflect the overall diversity. In some parts of the world, the composition of the invertebrate fauna of open waters is well known.

To overcome the sampling and identification problems related to exhaustive surveys, Williams and Gaston (1994) suggested that higher taxon richness may be a reliable predictor of total species richness. They showed this to be so for ferns and butterflies in Great Britain, for passerine birds in Australia and for bats in North and Central America. As identification to family is readily achieved and less time-consuming than identification to species, this is a useful tool where it applies.

A site that is rich in species is generally regarded to be of high conservation value. This is closely linked with diversity of habitat, water chemistry and the size of a site. However, it should be noted that the case for species richness depends on how many and which groups are included in the analysis: for example, an oligotrophic site may be very species-rich if desmids are included, a eutrophic site very poor if diatoms are examined. There is a tendency for the number of species of plants and animals to be positively correlated with the area of the site. This is not detrimental as larger sites are easier to conserve. Latitudinal (Pianka, 1966) and altitudinal differences influence the number of species possible and this should also be taken into account in assessments. At high latitudes and altitudes the number of species is limited by past glacial history and present low productivity and temperatures.

Eutrophic water bodies are usually rich in species and diversity counts high in their selection. Excessive nutrient enrichment in such water bodies, however, can lead to the well-known phenomenon of eutrophication, where intense algal blooms reduce light penetration, with the consequent death of macrophytes and the fauna that depends on them (Olsen, 1964; Morgan, 1970). Such sites are of little value unless the process can be reversed. On the other hand, oligotrophic sites are characterized by their paucity of habitats and species composition. Indeed, the finest sites for oligotrophy, extremely poor in nutrients, have a very low species diversity. This criterion must therefore be used with caution when examining such sites and the importance of paucity must be taken into account.

A prerequisite for maintaining maximum biological diversity in a region is to identify a reserve network that contains every possible species. Margules *et al.* (1988) examined all the plant species of 432 wetlands in Macleay Valley, Australia and constructed algorithms to determine the smallest number of sites to contain all species – in this case only 4.6% of the total. It should be noted that the factor of size came into play as these sites actually made up 44.9% of the total area.

(d) ***Extent***

As stated above, diversity increases with the size of a site. Generally, the larger a site the greater its viability, i.e. its resistance to changes brought about by internal or external influences increases. Clearly, a small pool on agricultural land is more vulnerable than a large lake covering several square kilometres. This vulnerability will also be influenced by the intrinsic fragility of the habitats represented. A site that contains the whole catchment of a water body is the most desirable from the practicality of protection, but in highly developed areas this is rarely possible. The larger a site the greater the potential for those species, particularly predators such as otters, that require a large home range. Larger sites that are equal on other grounds should take precedence over small sites.

(e) ***Rarity***

The occurrence of species, communities or habitats that are rare or outside their normal range of distribution adds to the importance of a site. Species existing at the edge of their range may present various research possibilities and be genetically distinct from other populations. Some species and communities are rare because they have specialized habitat requirements (e.g. coregonid fishes, which require unpolluted lakes). Many have greatly retracted their distribution as a result of human pressures, such as reservoir construction and pollution, which have made their habitat uninhabitable. Care must be taken not to apply too much weight to certain groups; for example, some species of birds, with their strong capacity for movement, can arrive as vagrants, nest and equally suddenly disappear a year or two afterwards, independent of the management applied. On the other hand, if a site is unique for certain species, as in the case of Lake Baikal (Anon., 1969) and the African Rift Valley lakes (Coulter, 1991), this adds strongly to the case for their conservation and evokes the concept of irreplaceability.

(f) ***Fragility***

This is a complex criterion involving both the intrinsic vulnerability of habitats and the viability of populations. Species with specific and narrow requirements are particularly vulnerable. For example, plant and animal populations of eutrophic habitats have a certain resistance to nutrient enrichment, whereas those of ombrogenous mires and oligotrophic water bodies are extremely sensitive to slight nutrient increase, being replaced then by other species. If the main interest of an area is in relation to its seral successions, which are dynamic features, it is important to decide whether it is possible to keep the stages represented by management. Fringe or relict species may be extremely difficult to maintain and their fragility may be such that their viability may be very doubtful even under favourable management.

The surface of certain peatlands, such as *Sphagnum* lawns, is extremely susceptible to regular trampling and the practicability of preventing such activities must be taken into account. The physical form of a site may contribute to its fragility. A long, linear and narrow blanket mire has a much higher perimeter-to-surface ratio than one of similar area that is round. It is consequently more vulnerable to detrimental activities on the surrounding land. Where possible, such sites should be avoided.

At the species level, the proportion of the world population that would be contained in a proposed protected area may be important. This relates to the degree of cohesion of a population and the continuity with or isolation from other protected areas. Sophisticated techniques of population viability analysis (PVA) are now being developed to determine the minimum area of individual reserves and the optimum spatial arrangement of reserve network necessary to maintain a species (Thomas *et al.*, 1990; Lamberson *et al.*, 1992). However, this technique can only be used when detailed information is available on such population parameters as rates of fecundity, mortality and dispersal, which at the moment is only available for a small number of species.

(g) Representativeness

The concept of representativeness is based on the premise that a protected natural area should contain the full range of natural variation characteristic of a landscape unit or region (Austin and Margules, 1986). Saetersdal and Birks (1993) carried out multivariate analyses to show the relationship between species assemblages and environmental and other variables for 60 deciduous woodlands in Norway. Two-way-indicator species analysis revealed four representative types, of which one was not represented in the existing reserve series.

The range of representative types is often related to climatic and edaphic factors (Belbin, 1993) and will therefore vary geographically. Where multivariate analysis has not been carried out for the habitats concerned, determination of representativeness will depend on other techniques and on the assessor's personal experience.

Where it is intended to select a series of reserves to show the complete range of ecological variation of fresh waters in a region or country, it is necessary to make reference to the classification of sites (sections 7.2 and 8.3 and Figures 1.6 and 4.2) to ensure that the plant and animal communities considered typical of this type of water body are represented. When selecting sites as reserves it is always tempting to select the unusual but it is equally important to include typical sites characteristic of each habitat, even though they may be commonplace. This gives the opportunity to be rigorous in applying other criteria when choosing very high-quality sites. Often, sites where other criteria are important may contain a wide range of characteristic and common species.

(h) Other criteria

The above criteria are probably the most important in site selection but varying situations and aims may modify the importance of a particular criterion. It is necessary to remain flexible in their application. They are summarized in the list above and should be considered as guidelines in any conservation programme.

Other supplementary criteria may also be applicable: a site with an extensive recorded history or where much research has been done can be of high importance for future studies even though it may not score highly on other criteria. The potential for educational purposes may carry great weight if the site is in proximity to large towns. Contiguity with other sites already protected for their conservation value provides an extra protection from external influences and provides the possibility of pooling management resources. A habitat containing species at the extreme of their geographical distribution is of interest, particularly as the potential for genetic change is high in these situations, where environmental pressures against survival are limiting.

Finally, the feasibility of effective conservation must be taken into account where issues such as industrial development or other detrimental activities, such as lowering the water table, exist or are about to take place. The lowering of the water table around the Norfolk Broads by agricultural drainage and shrinking of the peaty soils left the existing nature reserve of Calthorpe Broad perched above the water table so that water levels started to recede. This was only alleviated by the placing of an impermeable plastic apron around the reserve to retain water. This type of situation should be avoided wherever possible. Where a site of exceptional ecological value has been degraded the possibility of rehabilitation must be taken into account when deciding whether to protect the site. If restoration is not feasible then, generally speaking, the site should not be selected and an alternative site should be chosen.

4.2.4 Scoring

Various methods of scoring the value of a site to give more objectivity to the evaluation process have been tried (Blyth, 1983; Rabe and Savage, 1979). Unfortunately none of these are completely satisfactory (Usher, 1986). Attempts to weight criteria in relation to their importance have also not proved completely successful. For some well known groups (e.g. fish), simple analytical systems, which do not involve scoring, can be of value (Table 4.1).

Goldsmith (1983) differentiated between 'ecological (principal) criteria' such as extent, diversity and rarity, which could be more or less precisely measured, and 'conservation (supplementary) criteria', such as potential value and intrinsic appeal, which are value judgements and more subjective.

Table 4.1 Summary of a procedure for the selection of important sites for freshwater fish in the British Isles (after Maitland, 1985)

Question	*Answer*	*Check*
1. Does the site have fish?	Yes	2
	No	A
2. Is there sufficient information?	Yes	3
	No	B
3. Are there rare species?	Yes	C*
	No	4
4. Are there any unusual races?	Yes	C*
	No	5
5. Are any stocks pristine?	Yes	C*
	No	6
6. Are unusual species combinations present?	Yes	C*
	No	7
7. Is diversity high relative to the number of local native species?	Yes	C*
	No	D
8. Is the site a particularly useful representative, locally or nationally, of a particular type of aquatic habitat or fish community?	Yes	A
	No	D

Comments
A. If it is naturally fishless, or particularly representative, then it is of potential value and further attention should be given to its plants, invertebrates, amphibians and birds to see if they are of interest
B. Routine survey procedures should be carried out to assess the status of the fish community
C. The site should be considered as an important one and notified accordingly
D. The site is probably not of importance as far as freshwater fish are concerned
* The case for conservation is strengthened if C is reached by more than one route. Thus, for example, because Loch Lomond is relevant under five categories its conservation value is extremely high.

When reliable information is available for the number of species present, which is most likely to be possible for the macrophytes and vertebrates, a score can be given that relates to their actual numbers, but this will only be possible for the invertebrates in simple situations or where studies in depth have been carried out. Margules and Usher (1984) separated criteria in relation to the size of sites. For small sites ecological fragility, threat, species and habitat were most important and for large sites representativeness, size, naturalness and position in an ecological/geographical unit. If such a separation is real, there remains the problems of how one treats intermediate sites and where one draws the line among them.

An extensive botanical survey of 677 km of Welsh river corridors in order to determine the conservation interests is described by Slater *et al.*

(1987). They used three criteria: diversity, rarity and site uniqueness. Variation in the capability of the recorders and seasonal variation were the major causes of variation within the data. This had the greatest effect on the usefulness of the data to determine rarity. O'Keefe *et al.* (1987) describe a computerized system, used for the assessment of conservation status in South African rivers, which goes a long way to overcoming the problems associated with scoring. To arrive at a weighting for different groups of the biota and various attributes of the river and the catchment, they sent a questionnaire to river ecologists and conservationists throughout South Africa asking them to quantify the relative importance of the attributes. The weightings applied were based on the means of these replies which has the advantage of reducing the bias of individuals. Boon *et al.* (1994, 1996) have built on the method of O'Keefe to provide a sophisticated computerized scoring system, SERCON, to aid in assessing the conservation value of rivers in the United Kingdom.

In many scoring systems, scores (say 0–3) for each criterion are given in order to aid selection among a large number of sites, but they should not be regarded as absolute differences. The sum of the scores for the principal criteria (section 4.2.3) can be used to compare the ecological value of habitats of a similar kind (e.g. within a set of raised bogs, swamps, oligotrophic lakes, montane streams, etc.). The summed scores for supplementary criteria can then be used when it is necessary to make a choice between sites of similar ecological rating, but the principal criteria must take priority for the evaluation. Weighting of certain criteria thought to have a higher importance in selecting sites is a delicate matter, as the importance of a weighting factor is often an arbitrary decision. Usher (1986) suggests that the inter-relationships between criteria are too complex for any meaningful index to be formulated. Goodfellow and Peterken (1981) considered that indices disguise the real biological characteristics and make the judgements almost impossible to check. Smith and Theberge (1986) conclude that any method of evaluation must be seen simply as an aid to human judgement. The important thing is to make a standard approach to all sites of a particular type. Even with the most sophisticated statistical techniques a degree of subjectivity enters into decision making. Until truly objective assessments can be made, an oversophistication of approach is profitless and tends to give a false impression of objectivity. The value of scoring is that it aids the determination of the conservation value of sites in a standard and repeatable manner and thereby the selection of the highest value sites for protection.

4.2.5 Sites for aquatic birds

A special case must be made for the evaluation of sites for aquatic birds, as they are mobile and may use different sites for breeding and overwintering.

Waders, ducks and geese can gather in thousands at particular staging posts during migration to build up their energy reserves. They may only remain for a few days but this stop is essential to enable them to achieve the next leg of their migration. If the site is destroyed the future of this population may be threatened. Reichholf (1977) considers that, for waterfowl, these staging posts should be no more than 400 km apart. Thus, several sites along the migration route may be essential for the conservation of a population and the criteria discussed above for general biological values may not all be necessary.

To meet the needs of aquatic birds a set of criteria particularly directed to the requirements of migrating and wintering waterfowl were drawn up at a conference in Heiligenhafen, Germany, in order to select sites for protection under the Ramsar Convention (Anon., 1976). These criteria are divided into four groups, those pertaining directly to aquatic birds being given as absolute numbers or percentages of the population using the site. The fourth group makes the point that, notwithstanding the fitness of a site to be considered as internationally important on other criteria, it should only be included in the Ramsar list if it is physically and administratively capable of being effectively managed and free from the threat of a major impact. The Heiligenhafen criteria are specifically aimed at evaluating the bird populations using a site but do not cover the broader criteria necessary to determine the value of the habitats represented (Morgan, 1978).

The Heiligenhafen criteria are as follows:

1. **Criteria pertaining to a wetland's importance to populations and species** – A wetland should be considered internationally important if it:
   - (a) regularly supports 1% (being at least 100 individuals) of the flyway or biogeographical population of one species of waterfowl;
   - (b) regularly supports either 10 000 ducks, geese and swans; or 10 000 coots; or 20 000 waders;
   - (c) supports an appreciable number of an endangered species of plant or animal;
   - (d) is of special value for maintaining genetic and ecological diversity because of the quality and peculiarities of its flora and fauna; or
   - (e) plays a major role in its region as the habitat of plants and of aquatic and other animals of scientific or economic importance.
2. **Criteria concerned with the selection of representative or unique wetlands** – A wetland should be considered internationally important if it:
   - (a) is a representative example of a wetland community characteristic of its biogeographical region;
   - (b) exemplifies a critical stage or extreme in biological or hydromorphological processes; or
   - (c) is an integral part of a peculiar physical feature.

3. **Criteria concerned with the research, educational or recreational values of wetlands** – A wetland should be considered internationally important if it:
   (a) is outstandingly important, well situated and well equipped for scientific research and for education;
   (b) is well studied and documented over many years and with a continuing programme of research of high value, regularly published and contributed to by the scientific community; or
   (c) offers special opportunities for promoting public understanding and appreciation of wetlands, such as being open to people from several countries.
4. **Criteria concerned with the practicality of conservation and management** – Notwithstanding its fitness to be considered internationally important on one of the criteria set out under 1, 2 and 3 above, a wetland should only be designated for inclusion in the List of the Ramsar Convention if it:
   (a) is physically and administratively capable of being effectively conserved and managed; and
   (b) is free from the threat of major impact from external pollution, hydrological interference and land use or industrial practices.

   A wetland of national value only may nevertheless be considered of international importance if it forms a complex with another adjacent wetland of similar value across an international border.

Breeding populations of birds are not easily assessed. Ferdinand (1977) described a system that has been used successfully in Denmark, based on counts of indicator species and the definition of breeding numbers for different groups of birds that signify international, national and local importance.

The conservation of migratory fish, which cross international boundaries, poses the same problems and it is important to give adequate protection to a given population over the whole of its migratory range. For Atlantic salmon, *Salmo salar*, there has been considerable collaboration among the countries concerned to achieve this aim through the formation of the North Atlantic Salmon Conservation Organization (NASCO). There is, however, much need for international cooperation to conserve stocks of many other migratory species such as the sturgeon, *Acipenser sturio*, which is now almost extinct.

5

Protection and management

5.1 ESTABLISHMENT OF RESERVES

Once sites of high conservation value have been defined and selected, it is important to establish some form of protection for them. The premier step is to decide where the boundaries should be (Lomolino, 1994). Theoretically these should follow the watersheds of catchments to ensure control of the quality of all incoming water but this is rarely possible in practice, especially for the larger water bodies. Efforts should be made to acquire as wide a band as possible of the surrounding land within the proposed reserve or at least to designate such an area as a buffer zone. Adverse threats to the site will have been taken into account during the selection process and, where these are of such a magnitude as to seriously threaten the viability of a site, the site will have been rejected.

In developed countries, nature reserves are set up either by a statutory body responsible for protection or by voluntary associations. Generally the former takes control of the sites considered to be of international or national importance and the voluntary bodies control those of regional or local importance. In underdeveloped countries it is usually the state that administers the reserves. It is worth noting that in communist countries voluntary conservation organizations do not exist, while in western democracies they play a vital role in conservation work and act as a watchdog on government policy. Where sites are not already owned by the state, they are usually acquired by purchase, by leasing or by a legal agreement with the owners. Purchase is more satisfactory as it guarantees permanent control, whereas leasing and agreements are less binding and may be revoked after a given time. The laws and regulations vary considerably from one country to another but the important point is that the authorities should have the means, both administrative and financial, to carry out the necessary management and guardianship and that they should have access to sound scientific advice. The latter is normally provided by a committee of experts.

Without these components the conservation of any site is unlikely to succeed.

5.1.1 International protection

UNESCO has established two methods of protecting sites at the international level. Under the Man and the Biosphere programme (MAB), started in 1971, 'biosphere reserves' can be set up to protect representative areas of high conservation value (Anon., 1974). The aim is to provide a framework of protected areas at the intergovernmental level that combines conservation with development, research, education and training. Each reserve contains a 'core area' of natural or minimally disturbed habitats, which is completely protected. Around this is a 'buffer zone', which may be a multiple-use area. People constitute an essential component of this zone, with agricultural activities, settlements and other uses, and their activities are considered fundamental to the long-term conservation and sustainable use of the area. MAB national committees can put proposals for biosphere reserves direct to UNESCO without going through their governments. If fisheries are involved these must also pass through the Food and Agriculture Organization (FAO). To stimulate research UNESCO will make one-off payments for expensive apparatus for the reserve and pay for meetings and travel expenses to encourage experts to come to work at the reserve. In 1994 there were 324 biosphere reserves in 82 countries.

A second means of protection by UNESCO is the designation of 'World Heritage Site'. This is applied to architectural and cultural monuments as well as natural sites. It is not as well adapted to freshwater sites as the biosphere reserve as no money is available to promote research and proposals must come from governments. This is a distinct disadvantage when a water body, or its catchment, traverses the boundary of more than one country, as agreement of all countries must be obtained. Lake Tanganyika, a lake of immense conservation value, is a good example where, with four riparian countries, two French-speaking and two English-speaking, it is difficult to obtain a consensus of opinion and united action.

A means of protection specifically directed at wetlands and migratory waterfowl is provided by the Ramsar Convention (Anon., 1971). Each state that becomes a contracting party undertakes to designate important wetlands within its territory for inclusion in a list of wetlands of international importance in terms of ecology, botany, zoology, limnology or hydrology. Each undertakes to promote the conservation, management and wise use of migratory stocks of waterfowl and encourage research and the training of personnel in the fields of wetland research, management and wardening. Where a wetland is deleted from the list, in the national interest, the country should, as far as possible, compensate for any loss of wetland resources and should, in particular, create additional reserves for waterfowl and for the protection of an adequate portion of the original habitat. In 1996, the Ramsar list consisted of 808 sites in 93 countries. This convention therefore plays a very significant role in the conservation of important wetlands. Although originally strongly biased towards sites of high value to waterfowl, it is now broadening its scope to all types of wetland.

5.2 MANAGEMENT PLANS

All fresh waters are dynamic systems and subject to change. This is particularly marked for wetlands, where the rate of change can be relatively rapid, as from marshland to a terrestrial habitat by accumulation of organic matter, if the process is not controlled. Therefore, no matter what type of protection is applied, it is essential to prepare a management plan for the reserve. This serves to clarify the reasons for selecting the reserve, the aims and the mode of management to be applied. Even for a system that is evolving extremely slowly, such as an oligotrophic lake, it is important to have a plan, if only to define the constraints on management.

Where the whole of the catchment is within the boundary of the reserve, management may be relatively simple, but where most of the catchment is not protected, much of the management may be concerned with alleviating the effects of activities outside the reserve but within the catchment. The first part of the plan should be descriptive, to put together all the known information on the site.

5.2.1 Site agreement

The name of the proprietor, who may or may not be the managing organization, depending on the mode of acquisition, should be shown. Acquisition is usually by purchase, by renting or convention and details of the conditions of the acquisition, duration, constraints, date of the agreement, etc. should be given.

5.2.2 Description of the site

The location and extent of the reserve should be indicated on an accompanying map, with details of any outstanding features. This should be followed by a general ecological description of the area covering morphometry, climate, geology, hydrology, water chemistry, turnover time, flora, fauna and other relevant information. Details such as lists of flora and fauna recorded should be given as appendices and can be updated regularly.

5.2.3 Reasons for establishment

The particular qualities of the site, how it relates to similar sites and, if the site was acquired to protect it from some particular threat, the nature of that threat should all be recorded.

5.2.4 History

Historical records of the vegetation and land-use are extremely useful in determining future management, as the present-day situation is the culmination of all these determinants. They give clues as to cause and effect.

Palaeontological information is of value if the reserve is to be used for scientific or educational purposes.

5.2.5 Literature

A list of all available publications on the site complements the historical information. Inventories of the biota and reports of research carried out on the water body are particularly valuable.

The second part of the plan defines the actual management to be carried out.

5.2.6 Aim(s) of management

The clear definition of the aims of management is crucial for the future of the site. These will be linked to the reasons for establishment but supplementary aims such as the education of schoolchildren or the reception of the public may be added. The aims of management may be different from one part of the reserve to another. Thus in one area it may be to maintain the *status quo*, such as to keep the different stages of the seral succession, whereas in another it may be to return to an earlier stage or advance the rate of succession for the benefit of particular species or communities. Management is often designed to increase the numbers of certain plants and animals or, in the case of invasive non-endemic species such as *Fallopia japonica* and *Impatiens glandulifera*, which are spreading rapidly along river courses in Europe and certain parts of North America, to control or eradicate them. Aims of management are often preventative as, for example, in preventing the pollution of a river, the erosion of a lake shore or the over-exploitation of a fish stock.

It cannot be overemphasized that a clear definition of aims is essential for good management and a prerequisite for decisions on the methods of management.

5.2.7 Methods of management

It is not intended here to give a catalogue of the multiple methods of management applicable in freshwater reserves (Chapters 6-9 and Figure 5.1) but simply to indicate that in this section of the plan the methods of achieving the aims are defined saying how, where and when the work will be carried out.

Any constraints, such as the need to prevent deleterious activities in sensitive neighbouring areas (Figure 5.2) or damage to the water table, must be stated. Examples of management for different purposes are given later.

Figure 5.1 One of the many routine management tasks on a wetland nature reserve: ditch clearing on Loch Lomond National Nature Reserve.

Figure 5.2 A glaring example of unsympathetic engineering: the new outflow from Gartmorn Loch, a local nature reserve.

It is useful to draw up a recapitulative table, which should be attached as an annexe to the plan, setting out for each management activity:

- the work;
- the period when it should be done;
- the number of persons involved and their grade;
- the quantities of material and the machinery required;
- the costs.

This information can be entered on computer disks, which enables rapid updating and verification of the costs and progress of the work against available funding, power and materials.

5.2.8 Other activities

Research projects, which may be designed to refine management techniques or determine the functioning of the system, should be described here. Other research of a non-detrimental nature, where protection from outside influences is necessary, may be permitted in the reserve.

The reserve may also be used for educational activities, either by schoolteachers with their pupils, or as activities organized by the reserve warden. The visitor options may vary from visits to the reserve for the public to courses for university students in elementary limnology and wetland ecology. Essentially, all activities of this type must relate to the capacity of the habitat to receive them. Care must be taken not to damage the habitat or seriously deplete the numbers of any species. Some habitats, such as stony lake or stream shores, are resistant to trampling and could be used for activities like collection of macroinvertebrates providing they are not overdone. Collection of phytoplankton or zooplankton is normally not abusive. On the other hand, the surface of many types of peat bog is extremely fragile and will not resist damage by trampling. Disturbance of feeding or roosting waterfowl or other animals should obviously be avoided.

5.2.9 Guardianship

If it is important to protect parts of the reserve from intrusion by the public, or to control their activities elsewhere, it will be necessary to employ wardens to patrol the area. Their responsibilities must be set out in the management plan with a code of practice for dealing with the public in a positive manner. A work schedule should be given in an annexe.

5.2.10 Staffing

The names of permanent personnel who are responsible for the different activities in the reserve (management, reception of the public, education, guardianship, etc.) should be indicated in an annex, with a resumé of their functions. In small reserves operated by voluntary bodies duties will vary with the availability of individuals but the person with overall responsibility should be clearly indicated.

5.2.11 Authorship

Finally, the names of the authors of the plan should be given for future reference.

Each management plan should be reviewed and revised about every five years. Minor amendments in the intervening period can be inserted as an appendix and for this reason it is convenient to have the prescriptive part of the management plan in loose-leaf form.

5.3 MONITORING

An essential in any reserve is to measure or monitor certain parameters regularly, in order to detect changes with time. The repetition of measurements with time distinguishes monitoring from inventory or assessment.

Monitoring may be applied for a variety of purposes. MacDonald (1991) classified monitoring into seven categories for use in evaluating the effects of forestry activities on streams in the north-west of the USA. Four of these, which are quite distinct from each other and are applicable to nature reserves, are given here. They could equally well be applied to the monitoring of standing waters or wetlands.

5.3.1 Monitoring natural trends

Basically, this is the detection of changes that are quite independent of reserve management practices and are either natural or brought about by activities outside the reserve, which are usually out of the control of the managers (Barendrest *et al.*, 1995). Different techniques are used in different habitats (Goldsmith, 1994; Pollard and Yates, 1995). Measurements are made at regular, well-spaced intervals in order to determine the trends of particular parameters. These are often simple measures of water chemistry such as pH, alkalinity, phosphate and nitrate, taken in winter and summer, but those chosen will depend on the situation. Many biological parameters can be used. In wetlands these could be quadrats or transects of the vegetation. A simple measure of the invasion of a marsh by trees can be made by taking annual photographs from a fixed point. Sampling fish populations or counts of waterbirds are particularly sensitive methods of measuring the health of a site, since these groups top the food chain. If outside influences are suspected the method of monitoring will obviously be chosen to target these.

5.3.2 Implementation monitoring

Such monitoring is used to verify that a particular management or other activity has been carried out as planned. Normally this is carried out as an administrative review and does not involve water quality or biotic measurements. It will rarely be necessary except in the case of very complex management plans.

5.3.3 Effectiveness monitoring

This type of monitoring is designed to determine whether the management put into force is having the desired effect. The parameters will vary with the purpose of the management: they will show whether it has been successful, ineffective or negative and future management will be planned accordingly.

5.3.4 Project monitoring

This type of monitoring is related to specific projects proposed outside the reserve that may have an impact on the reserve. Thus the monitoring activity may take place entirely outside the reserve, depending on the type of project, or may also involve specialist measurements made within it. Examples of such projects are the construction of a dam upstream, discharge of effluent from a new factory or clear-felling of a forest in the catchment.

Trend monitoring and effectiveness monitoring are those which have the greatest application in nature reserves.

6 Wetland management

6.1 INTRODUCTION

Wetland management, under natural or semi-natural conditions, has generally been for hunting or nature conservation purposes, particularly in North America and Europe. Hunting management has been mostly directed towards duck shooting but other waterfowl and mammals can be involved. Although the aims may be different, the techniques are often the same for hunting and nature conservation, and management is aimed at maintenance or improvement of habitat, most frequently by manipulation of the vegetation and water levels. Marshes and swamps are more productive and dynamic than peatlands so that most management has been directed towards the former.

Peatlands are unproductive systems that evolve very slowly, on a time scale of thousands of years. In the acid, often anaerobic, conditions in the substrate, bacterial action is negligible and decomposition of the dying vegetation is slight so that it accumulates as peat. Nutrients are therefore locked in the system and the development of vegetation is severely limited, the characteristic plant species being adapted to these low-nutrient conditions. For continued growth the peat must remain saturated, either with groundwater, in soligenous systems, or with rainwater, in ombrogenous systems. The maintenance of a peatland does not usually involve direct management but protection from external influences may be necessary. Exploitation of peat on a commercial scale for fuel, as in Ireland and Russia, or loss by drainage for forestry, as in Finland (Goodwillie, 1980), not only devastates large areas of peatland but can lower the water table of neighbouring unexploited peatland. Similarly, adjacent agricultural drainage can draw off essential water, particularly from soligenous peatlands. The other great danger from agriculture is the arrival of wind-blown fertilizers or lime, which may permit non-characteristic plant species to establish in the richer nutrient conditions. Drift from herbicide applications can endanger certain components of the vegetation.

On the other hand, fens and marshes are more dynamic systems, richer in nutrients, which can progress rapidly from one stage of seral succession to another and eventually transform into a terrestrial habitat, with bushes and trees. According to the needs of management this often requires interventions in the reserve itself to maintain the vegetation in a particular condition, to open up pools, return to an earlier stage of succession, etc. The latter is the most common aim as the early stages have a higher biodiversity. As with peatlands, fens and marshes are vulnerable to drainage and vast areas have been lost for conversion into farmland.

Thus the basic requirement for wetlands is, wherever possible, to maintain the integrity of the hydrological system. This may mean some control, either by legislation or negotiation, of water withdrawal, retention, discharge or other use, outside the reserve boundaries. Management may be expensive, involving the construction of dykes and sluices and installing pumps to control the water level. Methods will vary in relation to climate, geomorphology, hydrology, soil type, flora, fauna and other factors. Those that have been found to be successful in the temperate zones of the world will not necessarily succeed in the tropics and field trials should be carried out before adopting these techniques on a large scale. Much management has developed by trial and error or simply by common sense, and will be refined as research deciphers the mechanisms involved. Management for individual species often involves the creation of very specific conditions for feeding, breeding or nesting. One of the aims of management should be to promote various habitats, which permit mobility of species between them as well as higher biodiversity (Segerstrom *et al.*, 1994).

The problem of drainage due to activities outside the wetland will vary with the individual case, will often require a custom-built solution adapted to the situation and will therefore not be dealt with in the present general account.

6.2 MANAGEMENT TECHNIQUES

6.2.1 Water regime

The seasonal, or more erratic fluctuations in water level that occur in many wetlands strongly influence the plant communities present. Most submerged species of true aquatics cannot resist a period of drying out, particularly during the summer, whereas emergent species, such as *Phragmites* spp., *Typha* spp. and *Schoenoplectus* spp., are more tolerant. Other species, such as *Polygonum amphibium, Littorella uniflora* and *Ranunculus baudotii*, are adapted to an amphibious life and a host of annuals invade wetlands during periods of low water levels. Drawdown of water level to expose the surface of the substrate is essential for the germination of some emergent species, such as the ubiquitous *Phragmites australis*. The speed of drawdown is important. Slow drawdown, over 2 or more weeks, produces conditions favourable for a wide variety of plant species, which develop at different stages. Rapid lowering of water level, within a few days, produces similar conditions over a wide area of the wetland at the same time and tends to lower plant diversity. Whether slow or fast, the seed production of emergent plants is generally higher when drawdown takes place in the spring, whereas late drawdowns result in higher stem densities and greater species diversity (Frederickson and Taylor, 1982). For management purposes, sluices are often operated on marshlands so that the water levels can be manipulated to produce the plant communities desired.

The two most important factors that determine the response of aquatic plants are the timing of annual drawdowns and the stage of the seral succession. Frederickson and Taylor (1982) found in experimental ponds in the USA that early drawdowns tended to stimulate the germination of *Polygonum* species on sites that were at an early successional stage, but these species were less likely to respond to drawdown by the third year of vegetation development or after continuous flooding. Mid-season drawdowns resulted in millets and late-season drawdowns in spangle-top, *Bidens* spp., *Panicum* spp. and crabgrass. Once areas have been under wet soil management for 4 years or more there is a gradual increase in perennial species. Annual drainage of some natural marshes, over an extended period, may cause nutrient impoverishment due to runoff and leaching. On the other hand, extended periods of flooding (e.g. 3 or 4 years without draining) can cause the locking up of nutrients in the organic sediments. These are only released in an available inorganic form when the marsh dries out. Nitrogen is the nutrient most likely to become limiting in this way. The phenomenon is well known in fish culture and some Central European fishponds are drained and put under an agricultural crop every 3–5 years, to mineralize the organic sink to an available form.

Although different plant species will be present in different geographical and climatological zones, the timing, speed, depth and duration of inundation should be varied to achieve the desired effect on the vegetation, as the principles of management probably remain the same. Refined management of wetland plants, in the semi-natural conditions that often exist in wetlands today, involves precise control of water levels and the timing of changes by means of sluices and by pumping of water where necessary. To increase diversity, larger marshlands can be divided into separate basins by means of dykes or natural features and each basin kept at a different stage of development, so that one may be dry when another is flooded.

In fens and marshland where the vegetated surface has mounted above the water table, either by natural succession or where the water table has been lowered by drainage, the wetland can be regenerated by digging out the substrate with a mechanical digger. Care should be taken to avoid undue compression of the soil. A series of cuts to different depths below the water surface, say between 20 and 80 cm, will enhance diversity. This allows colonization by pioneer species such as *Typha angustifolia* and *Cladium mariscus*.

Changing the water regime in intact peatlands is rarely necessary or desirable. Only where the water supply has been diverted from soligenous mires, usually by agricultural or forestry drainage, will it be necessary to attempt to re-establish the water supply, but care must be taken to see that its quality has not changed. Recently, interest has focused on attempts to revive the peat-forming process in areas where the peat has been extracted, by rewetting the remaining peat and encouraging the establishment of mire vegetation (Nick, 1984; Egglesmann, 1987). Fundamental difficulties are: obtaining water low in nutrients, preventing its loss where the peat has been cut through to underlying porous soils and stopping enrichment with inorganic ions from these other soils. These conditions may lead to other plant communities, such as fen. In peatland it is generally only necessary to maintain a constant water level and variable height sluices are therefore not necessary. A simple and aesthetically pleasing barrier can be built in the drainage channels out of peat excavated elsewhere on the site (Figure 6.1). It is necessary to protect this from erosion with a covering of stout butyl sheet of the type used for making agricultural ponds and to have an effective apron on the downstream side. The overflow should also be lined with the butyl sheet and the structure covered with peat turfs and bound together by planting peatland shrubs such as *Myrica* and *Vaccinium* spp.

Meade (1992) flooded an old peat cutting in England where *Molinia caerulea* and *Betula* had established, by raising the water level by 50 cm. After 14 years, *Sphagnum cuspidatum* had developed on the tufts of *Molinia* which were not completely submerged and on the *Juncus* which had developed around the water's edge. In deeper water where the *Molinia* was completely submerged it had died and hummocks of *Sphagnum fimbriatum* had taken over, with patches of *Drepanocladus fluitans*. Meade considers that the tussocks of *Molinia* and *Juncus*, which often develop in these situations, form an important support for *S. cuspidatum* and that waterlogging of existing tussocks, rather than total inundation, is the more effective principle to employ for raised bog rehabilitation.

Etherington (1983) regenerated a peatbog in Wales by stripping off the *Molinia* covering of small plots, which were then colonized by *Sphagnum* spp., *Drosera rotundifolia, Rhynchospora alba* and *Lycopodiella inundata*. More refined contouring of the surface to produce pools, depressions and hummocks aids diversity and produces habitat for aquatic insects specific to bog pools, such as certain dragonfly species. There should be a firm moist layer of peat 50–100 cm deep below the cut, providing a clean impermeable surface. Soupy peat surfaces are not easily colonized by plants. In such

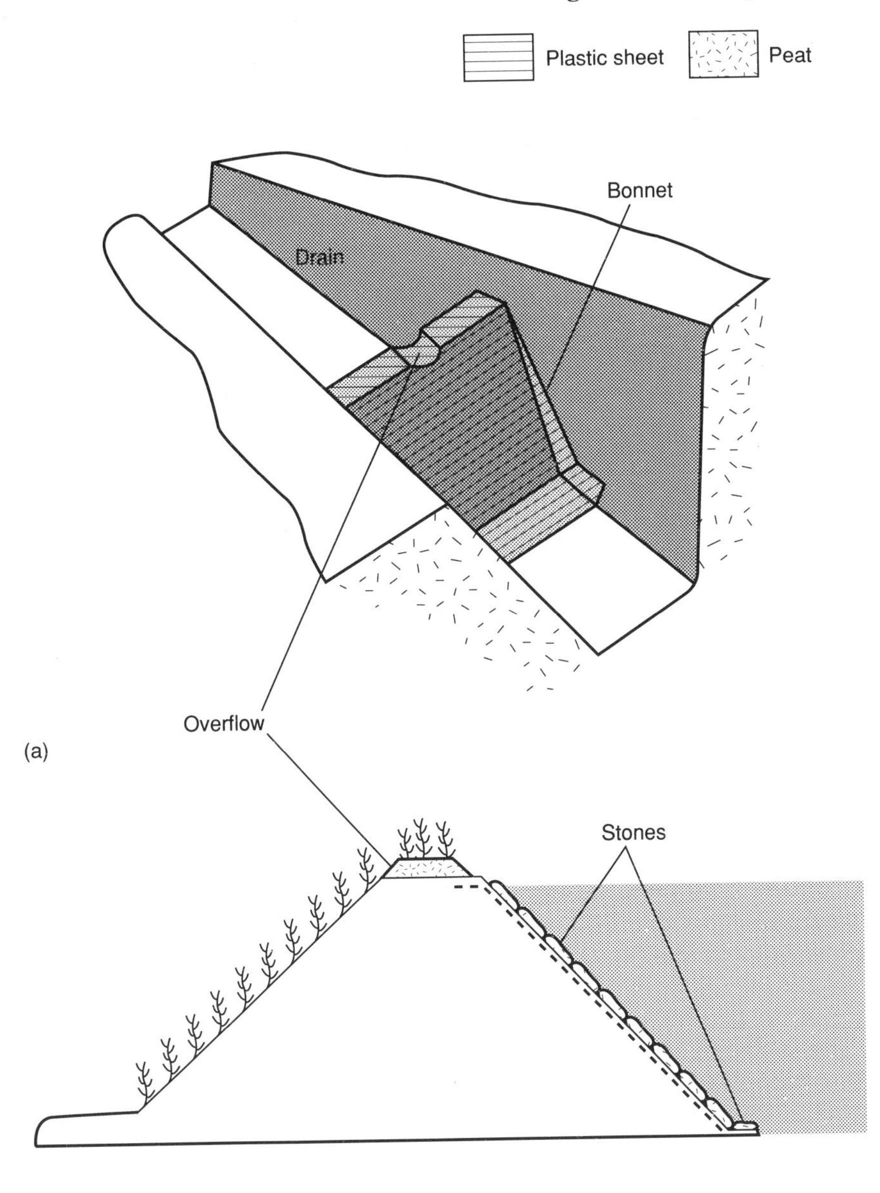

Figure 6.1 A peat dam reinforced with butyl sheeting. (a) General view. (b) Longitudinal section.

operations it should be remembered that too much manipulation destroys the stratigraphic record. It should be emphasized that re-establishment of peatland vegetation is an extremely slow process, as evidenced by a raised bog in France, at an altitude of 900 m, where the peat surface was exposed by the passage of a four-wheel-drive vehicle. Eight years later there is no trace of vegetation on the bare peat surface.

Senescent peatlands are often invaded by trees such as *Betula, Pinus* and *Salix* spp. To maintain the nature of the peatland, young trees can be killed by raising the water level but it is necessary to cut older trees near to the ground. The stumps should be treated with a herbicide such as glycophosphate, which should be brushed on to the stump to prevent damage to the surrounding vegetation (Cooke, 1986).

6.2.2 Grazing

In order to maintain the biological richness of many wetlands, management is required to prevent succession towards homogenous reed beds or drier situations with eventual invasion of woodland. The control of emergent vegetation, such as *Phragmites* and *Typha*, to prevent its extension or to create open water conditions for submerged aquatics such as Characeae and *Potamogeton* spp., is frequently carried out by grazing with cattle or horses (Chabreck, 1968; Bakker, 1978; Duncan and d'Herbes, 1982). Horses (Figure 6.2) are capable of grazing down tall *Phragmites* completely, whereas cattle prefer the young growing stems.

Also, horses consume twice as much plant material as cattle, and thus a lower density of horses achieves the same grazing impact. Heavy grazing by medium-sized horses, at 90 horse days per hectare, can eliminate dense swards of *Phragmites australis*. *Scirpus maritimus* is also readily eaten and the height of the stem greatly reduced, but its density is not reduced as the meristem is below the ground. Other species, such as *Typha angustifolia* and *Scirpus littoralis*, are only palatable to cattle and horses when they are young, so that the grazers must be herded on to them at the beginning of the growing season. Thus, if not carefully managed, grazing on emergent vegetation can be selective and lead to the spread of certain species such as *Typha*, which propagates rapidly vegetatively. Electric fencing is a useful management tool, facilitated by the use of simple solar captors to generate the electricity, which is easily moved to control the position of grazing animals. Permanent fencing should be used to protect areas where grazing is not desired. Cattle and horses can kill some trees, such as willows and poplars, by stripping the bark, and prevent the establishment of trees in wet areas where they graze (Lecomte and Leneveu, 1986). Seed production is reduced in grazed areas owing to destruction of the flower heads or severe reduction of the vegetative parts of the plants. In the Gramineae, which are well adapted to grazing, this is compensated for by an increase in vegetative

Figure 6.2 Marsh management using grazing by horses at Pagney sur Meuse, France.

reproduction. For most species, grazing decreases the rate of growth and survival but increases the recruitment of young plants, except under very heavy grazing pressure (Gordon *et al.*, 1990).

The trampling action of domestic herbivores, particularly at high densities, has a marked effect on vegetation (Haslam, 1972a). The rhizomes and root systems are broken up and this may kill certain species. Trampling of the soil can cause compacting and in clay soils puddling of the surface can cause prolonged suspension of the clay particles, reducing light penetration and the growth of submerged macrophytes.

The recycling of plant material in the dung of animals can increase the productivity of wetlands (Fox, 1976), and Granval (1988) found that the density of lumbricoid worms was twice as high in grazed areas as in mowed areas.

Fiala and Kvet (1971) found that greylag geese, *Anser anser*, had marked grazing effects that could temporarily eradicate *Phragmites* along the shores of Czechoslovakian fishponds. The muskrat, *Ondatra zibethicus*, has been used in North America to manage emergent vegetation and can open up dense stands of *Typha*. As a management tool it has the disadvantage that when numbers increase it must be controlled by trapping, otherwise the increased numbers cause overgrazing and perforate dykes with their burrows. Control is economically practical with this species as the pelts are valued for the fur trade (Bishop *et al.*, 1979). The use of wild grazers to manage vegetation is attractive in areas where they are endemic, provided their numbers can be manipulated. They should not, however, be introduced into other areas for this purpose, as this may result in population explosions, with devastating effects. In the Norfolk Broads, England, the introduction of the coypu, *Myocastor coypus*, caused an enormous decline in *Phragmites* and *Typha* spp. although there was little effect on other species such as *Acorus calamus* and *Iris pseudacorus* (Ellis, 1963).

The control of emergent vegetation can also be achieved by mechanical cutting and mowing, which can be more precisely controlled. Large wild mammals have been eliminated from many wetlands by man and, in the absence of these, the employment of domestic stock to graze vegetation restores a 'natural' component to the ecosystem. This not only has an aesthetic appeal but the grazers produce a usable crop in the form of meat or marketable young horses for riding. This may make this method of management more economic than mechanical cutting. The rustic and hardy Highland cattle have been introduced from Scotland into a number of wetlands in continental Europe, not only on account of their hardiness but because their primitive aspect appeals to people.

6.2.3 Mowing

The control of emergent vegetation by cutting with mechanical cutters attached to boats, amphibious vehicles or tractors has been widely practised. As one can be more precise with mowing than with grazing, it may be preferable to use this method in small marshes, where precision may be important and where the husbandry of livestock may not be practical because of lack of space. The control of *Phragmites* is more effective when they are cut under water, in the spring, after which water enters the cut ends of the stems and causes them to rot. For rapid control of *Phragmites*, Björk (1976) recommends the cutting up of the rhizomes with a rotovator and permanently flooding the area afterwards. The pieces of *Phragmites* float to the surface and are carried to the water's edge where they can be gathered up. *Phalaris arundinacea* and *Typha latifolia* have been controlled in ponds at Münster, Germany by ploughing down to a depth of 45 cm, which goes well below the root system, followed by flooding to prevent the germination of seeds (Anon., 1982). Dykyjova and Kvet (1978) recommended that, in areas with cold winters, ponds should be drained in winter to kill *Typha* by freezing the rhizomes, or alternatively, drying them out during hot summers.

On the other hand, mowing may be used to favour the growth of *Phragmites*, where marsh vegetation has been degraded by drainage or eutrophication. Gryseels (1989a) cut plots during winter in a Belgian marsh, where the reeds had been completely overgrown by competitive vegetation such as

Urtica dioica and *Calystegia sepium*, for five successive years. This resulted in the rehabilitation of the reed bed, which became more uniform, the *Phragmites* produced flower heads, which it had not done before management, and the competitive vegetation disappeared. Gryseels considered this to be the combined effect of cutting and litter removal. Cutting must continue each winter as long as the conditions that allowed the invasion of other species persist. In the same marsh, successive summer cutting led to a decline in *Phragmites* and the development of a diverse hayfield vegetation (Gryseels, 1989b).

Submerged vegetation has frequently been controlled in wetlands and shallow lakes for navigation and recreational purposes. This should not normally be necessary in conservation areas, as these plants are generally considered as an asset in terms of the aquatic invertebrates they house and nourish, as spawning grounds for fish and amphibians, and as a food for vegetarian aquatic birds such as coot and swans. Stewart and Davies (1986) have shown that cutting *Potamogeton pectinatus* in a South African lagoon reduced the biomass of benthic invertebrates significantly.

6.2.4 Burning

Controlled burning of dry emergent vegetation has been used as an effective management tool, especially in North America (Linde, 1969). The principal uses are:

- to remove dead vegetation and litter, to avoid the build up of debris on the marsh floor;
- by means of hot fires to burn down into the organic substrate and create depressions in the lake floor;
- to produce a succession of grasses and herbaceous plants to provide feeding and nesting areas for waterfowl;
- to burn out invasive shrubs and saplings.

Burning into the lake floor to produce depressions is only applicable when the marsh has an organic peaty substrate that will allow the fire to burn downwards. This can only be done under still air conditions, when the fire is slow-moving and can become very hot. Generally, burns of this type produce depressions up to 50 cm deep but if the fire gets well into the underlying peat potholes up to 2 m deep can result (Linde, 1969). Burning is a cheap way of producing an interspersion of shallow pools throughout an area with dense emergent vegetation, but care must be taken to control the burns so that they do not run wild into neighbouring areas. Burns for this purpose are usually carried out during the dormant season. Extensive burning to remove plant debris is best carried out in the autumn, when the vegetation is still green and the fire is more easily kept in hand. Burns to control shrubs produce the best results when shrub growth is still active in late summer. It goes without saying that control of vegetation, by whatever means, should never be carried out during periods when marsh animals are reproducing or when young animals are still dependent on plant cover.

Haslam (1972b) found that an early burn of *Phragmites* at the beginning of the winter gave earlier growth and more dense production of shoots in spring. On the other hand, late-winter burns damage emergent buds and growth is retarded as replacement shoots take 2–4 weeks to come up. Burning of the basal litter exposes the soil to greater fluctuations in temperature and can be harmful to the fauna, particularly invertebrates (Haslam, 1973). Shay (1984) carried out controlled experiments on 20 plots in a *Phragmites* marsh in Manitoba, to determine the effects of the timing of burning. Burning was carried out once in August, October and May of successive years. The results showed that summer burning considerably reduced regrowth but that spring and autumn burns increased it. The August-burned plots suffered an initial reduction in biomass compared with the controls but 3 years later had returned to the same level, whereas the biomass of the spring- and autumn-burned plots was still significantly higher. Shay postulates several possible reasons for these events. Shoot density initially increases because fire stimulates the sprouting of lateral buds away from the apex of rhizomes. Spring and autumn burns stimulate productivity by reducing litter, which may have a biomass greater than the living shoots (Thompson, 1982), allowing the entry of more light and air. Fire-darkened soil warms up more quickly than unburnt litter and burning releases nutrients. When a stand is burned in summer there is enough time for a secondary regrowth of shoots before they are killed by frosts. This depletes rhizome reserves and substantially reduces growth the following year.

Ahlgren (1963) found that burning increased soil fertility temporarily in the top few centimetres of soil. Nitrogen, phosphates, magnesium, potassium and calcium increased immediately after the fire and nitrogen, potassium and calcium returned to preburn levels during the second growing season. Other studies have shown that grasses arising from a burned area are more nutritious, have more minerals and protein, and are more tender than grasses in unburned areas (Komarek, 1965).

6.2.5 Herbicides

When other methods of eliminating undesirable vegetation prove impractical it may be necessary to use herbicides but this should only be considered as a last resort. Experimental studies (Newman, 1967; Newbold, 1975) have shown that the decaying vegetation can completely deoxygenate the water, produce high levels of bacteria and kill out fish and invertebrate life. As water is an efficient transporter of herbicides it is important to see that they do not arrive in the vicinity of non-target plant species, and indeed the effects of herbicides on many species are not yet known. Where decomposition is slow, mats of filamentous algae may form. These restrict the macrobenthos that can live there to forms such as chironomids and oligochaetes, which can burrow in the mats of algae; these mats may persist for up to 2 years. Whatever herbicide is used it should be selective for the target species and of short half-life (Newbold, 1984). The method of application is important. Spraying is suitable for emergent vegetation provided it is done under calm air conditions, to prevent drift outside the treatment area, but for submerged plants application in a granular form is more desirable, to prevent dispersion by water currents. Where spot treatments cannot be carried out, it is recommended that only small areas of vegetation should be treated at one time and that 2–3 weeks should lapse between the treatment of neighbouring areas. The longer the phytotoxic period the greater the chance that other resistant plant species will establish themselves and massive growths of undesirable algae may develop. Herbicide treatment of emergent vegetation is more efficient if the area is mowed or burnt before application (Martin *et al.*, 1957). This removes dead plant material and allows better penetration of the spray into the living stems. It goes without saying that the recommendations of the manufacturers regarding dose rates for a particular use should be followed carefully. Formulations are continually changing and an overdose can have a negative effect by burning the leaves and reducing the amount of the chemical taken up by the plant.

Herbicide treatment has also been used for the control of undesirable algae. Bungenberg de Jong (1968) found that diuron was effective for controlling filamentous algae in fish ponds. Blue-green algae disappeared whereas desmids and diatoms appeared to be resistant. In many countries the use of herbicides in aquatic systems is restricted to well-tested, less noxious chemicals. In Sweden the use of herbicides is not allowed in protected areas as part of a general policy to reduce pesticide use (Larsson, 1982). According to Thomas (1982), dalapon and glycophosphate are the only two herbicides in regular use in the UK to kill grasses and broad-leaved herbs, and 2,4,5-T is used to control shrubby growth and prevent carr development. The latter has been applied extensively for shrub control in the United States (Linde, 1969).

As well as herbicides that are absorbed by the leaves, chemicals such as atrazine and simazine, which are absorbed on the soil, have been used in the United States. These persist up to 3 years in the sediments, compared with a maximum of 3 months for those named above, and have been used to maintain open areas free of macrophytes. Their use in a nature reserve would be very inadvisable.

The effect of herbicides on the invertebrates is generally an initial loss of herbivores, due to the loss of their food supply, but these recolonize if suitable replacement plant species establish themselves (Newbold, 1975). Working from experiments with diquat, Newbold emphasizes:

- the importance of using a selective herbicide allowing non-susceptible plant species to colonize and provide a new habitat for some of the invertebrates;
- the need for suitable unsprayed areas in the vicinity from which recolonization can occur;
- spraying should be carried out for one season only.

In this way habitat destruction may only be temporary but it could be more destructive depending upon the type of herbicide, the treatment level, the frequency of spraying and the suitability of any replacement plant species as a habitat for the fauna.

6.2.6 Dredging

In shallow lakes and wetlands problems of infilling may exist due to sedimentation of inorganic material carried in by inflowing streams or the rapid accumulation of organic matter produced within the lake as a result of eutrophication. The latter sediments are rich in nutrients such as phosphates, which may be released and contribute to nuisance algal blooms, which may shade out macrophytes, eliminate certain invertebrate and fish species (Morgan, 1970), be toxic to livestock or produce unpleasant odours on decaying. Deoxygenation, during hours of darkness, may cause summer or winter kills of fish. Infilling of lakes may also be caused by excessive growths of submerged or emergent vegetation.

Though sometimes costly, one way of removing unwanted sediments or macrophytic vegetation is by dredging. In artificial lakes, which can be drained to make access for equipment, and for areas close to the shore in natural wetlands this can be done using conventional draglines but for work farther from shore in permanent waters various floating hydraulic dredges have been used. Pierce (1970) describes techniques and materials used in dredging operations in 49 lakes in the Great Lakes region of North America. He found that 'cutterhead dredges', based on the design of larger dredges used in marine dredging, were used almost exclusively in these lakes. They are usually mounted on pontoons and the extracted material is pumped ashore through pipes 6–14 in (15–35.5 cm) in diameter. The pipes are attached to the hydraulically driven cutterhead, which has three to six blades in a spiral design that rotate and cut free the sediment. By means of two stilts, which penetrate into the lake bottom, the pontoon 'walks' forwards, while the cutterhead is manoeuvred from side to side in the sediment. The disposal of the dredged material must be considered carefully. Rich organic sediments may be profitably spread on farmland. On the other hand, inorganic sediments lend themselves to the construction of islands, which can be created directly during the dredging process (see section 6.3.4 for their design).

6.2.7 Rafts

Where it is not possible to construct islands to provide nesting habitat for aquatic birds this can be provided to a limited extent by the use of rafts anchored in open water (see section 9.3.3).

6.2.8 Explosives

Small ponds may be created in wetland sites by the use of explosives. This is usually done to produce a mosaic of ponds in wetland vegetation, providing habitat for wildfowl but also of obvious value to aquatic invertebrates. Whitman (1982) gives an account of the methods used by the Canadian Wildlife Service in New Brunswick. Best results were obtained using six to eight 23 kg charges of ammonium nitrate, mixed with fuel oil, placed in a crescent or circular pattern in firm soil about 1 m below the surface in holes 20 cm in diameter. The charges, placed at 5 m intervals, were composed of a 23 kg bag of ammonium nitrate with a stick of 40% ditching dynamite in the centre. The charges were connected in series by high explosive primer cord, which, when detonated by an electrical cap, resulted in simultaneous explosion of all the charges and the creation of ponds 25–30 m in diameter. Winter operation was considered to be the most efficient, because of the relative ease with which the materials could be carried on the frozen marsh. Of a variety of sites tested, peat substrates were the least successful; organic matter from the sides and bottom of the ponds rapidly floated over the surface, completely filling the open water area in a few days.

Advantages of using ammonium nitrate fuel mixtures are that it is safe to handle and store and the costs are considerably lower than other methods of pond construction (Linde, 1969). If funds are limited this may be the only way of creating open water areas in a wetland. It produces a more natural-shaped water body than those dug with a bulldozer or drag-line, which tend to be more rectilinear for ease of working. However, it is more difficult to predict size and depth in advance because of great variation in soil characteristics from place to place. Pond depths may vary from 30 cm to 2 m. A further disadvantage is that the edges of blasted ponds are often extremely steep, although it is not too difficult to ameliorate this by hand while the edges are loose after blasting.

Dramatic methods like this must be used with great care and with conservation as a primary purpose. Many beautiful bogs in Canada and elsewhere have been severely damaged by the indiscriminate production of open pools purely for wildfowling.

Figure 6.3 Aerial view of the Okavango Delta.

6.3 EXAMPLES OF MANAGEMENT

6.3.1 Marshlands

Areas of high ecological importance are often under threat from the local human population, which wishes to exploit them to meet the agricultural, industrial or domestic demands of an increasing population. The solution may be in a compromise based on effective forward planning, although it may not always be possible to integrate external demands with conservation. The possibilities should always be investigated and not dismissed out of hand.

The importance of long-term studies in determining whether, or how, the exploitation of resources can take place is well illustrated by the work of Ellery and McCarthy (1994). Various proposals have been put forward to take water from the Okavango Delta (Figure 6.3), the world's largest delta and a wetland of high international importance, for diamond mining, irrigation and drinking water (Anon., 1993; Scudder *et al.*, 1993).

This is one of the few sources of permanent water in an otherwise arid area. Ellery and McCarthy studied the functioning of the system and the changes that have taken place in the form of the inland delta. Both lakes and rivers have been formed and subsequently disappeared (Figure 6.4), radically changing the drainage pattern.

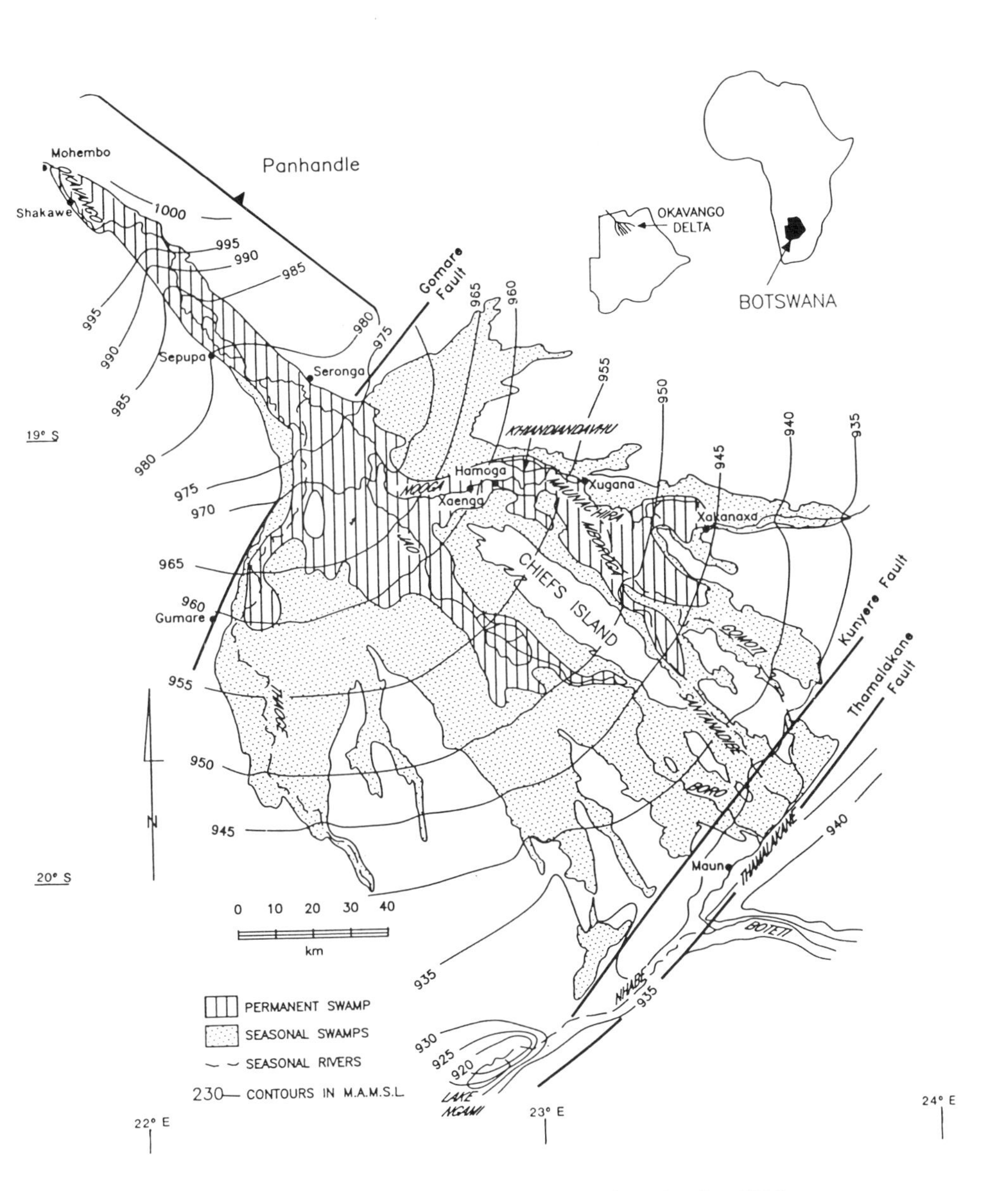

Figure 6.4 Map of the Okavango delta (after Ellery and McCarthy, 1994).

Ellery and McCarthy found that sediment introduced into the system by the Okavango River results in constant changes in the distribution of water on the fan surface. This, combined with the growth of aquatic vegetation, blocks existing channels, redistributes water and leads to a mosaic of different stages of wetting and accounts for the overall habitat diversity of the system. Some 96% of the water entering the delta is lost by transpiration, a high proportion of dissolved salts being fixed in insoluble forms in the soil, particularly on the islands. This prevents the development of saline lagoons. The temporal changes in the distribution of water over the delta surface promotes regeneration of saline soils that are locally toxic on islands. The key to the functioning of the system lies in the major loss of water from the system being by transpiration and not by evaporation, and the dynamic changes in water distribution brought about by the sedimentation in the upper delta. Thus any exploitation of the system by man must not upset this structure. Ellery and McCarthy believe that the removal of a small percentage of the inflow water would have a negligible effect on the system as a whole, in view of the large amounts available in the inflow river and changes in the area of permanent and seasonal flooding in relation to annual fluctuations in inflow. The method of abstraction would be critical. A dam situated on the inflow would interrupt the entry of sediment essential to the dynamics of the delta. A dam situated on the outflow would not only cause local salination problems but would soon be rendered useless by changes in the position of the outflowing river. They conclude that pumping of water from the inflow, in phase with the flood cycle but without the construction of a dam, would provide the needs of man and have the minimum effect on the present functioning of the ecosystem. In relation to any subsequent agricultural developments it would be important to prevent increased cattle grazing in the delta itself as salination of the surface water may occur if the vegetation cover were substantially reduced for a long period.

At the beginning of the century, Lake Hornborga in southern Sweden was considered to be one of the most important lakes for breeding and migrating waterfowl in Western Europe. It was particularly important as a staging area for thousands of migrating cranes, *Grus grus*. The lake covered an area of about 30 km^2 and had a maximum depth of 3 m. Successive lowerings of lake level between 1802 and 1933 to claim land for agriculture resulted in the lake drying out completely during summer. The construction of a dyke in 1954 led to up to a maximum of 80 cm of water resting in the lake basin in summer. The changes resulted in the fringing emergent vegetation, particularly *Phragmites australis* and *Carex acuta*, invading a large part of the lake surface (*ca* 12 km^2). Consequently, the numbers of waterfowl using the lake declined drastically. Additionally, the drained land proved inadequate for agriculture.

In 1967 the Swedish Environmental Protection Board commissioned studies to determine whether Hornborga could be restored as a bird lake. The work was facilitated by the use of Seiga amphibious vehicles (Figure 6.5), which have large balloon tyres allowing them to float and thus pass over liquid mud.

Figure 6.5 Cutting *Phragmites australis* at Lake Hornborga, Sweden, using a mower mounted on a Seiga amphibious vehicle.

Various hydraulically operated tools, such as mowing blades, rotovators and mechanical diggers, can be mounted on the platform of the Seiga. Using these, combined with subsequently raising the water levels to a maximum depth of about 2 m, areas of *Phragmites* were converted to open water (Björk, 1972). The dry stems were cut in winter and burnt on the frozen ground. In spring and summer the amphibious vehicles were used to cut the growing stems and then to rotovate the rhizomes so that they were cut into small pieces (Figure 6.6).

Large quantities of rhizomes, plus other plant material, were loosened from the bottom and transported to the shore by wind action in springtime, when water levels were high. This material dried out when water levels fell and was burnt in summer. The time for this operation was 8–10 machine hours per hectare (Björk, 1972).

Submerged vegetation of *Chara, Potamogeton* and *Myriophyllum* quickly established in the area formerly covered with reeds, with a rich benthos. Where *Carex* dominated the root mat it was too thick to be cut by the rotovator, but when the water level was raised this floated away from the bottom and could then be transformed into a mosaic of islands and channels using the digger. To recreate mineral shores it is important to remove *Phragmites* along shorelines that are orientated so as to be exposed to wave action, which washes away organic sediments and provides suitable

Figure 6.6 An area of *Phragmites australis* at Lake Hornborga, Sweden, that has been rotovated to reduce it in size.

habitat for wading birds. Ducks and grebes increased rapidly during the restoration. Björk emphasized the importance of maintaining the flow of water through the lake and allowing the free movement of ice, with its eroding effect, to keep the open lacustrine conditions in the centre of the wetland.

Similarly, Lake Hula, in Northern Israel, has been drained for agricultural purposes. Fed by the River Jordan and springs, it was the only freshwater lake in Israel and the last aquatic outpost facing the Saharo-Arabian desert belt. The flora and fauna, containing elements of the Ethiopian and Palaearctic biota, was extremely rich and diverse. Efforts to drain the lake and extensive marshes commenced in 1887 and the main drainage was completed in 1958. Efforts were made to retain a small part as a nature reserve, which covered 795 acres (322 ha) of wetland containing 270 acres (109 ha) of open water. The spring water entering the reserve was greatly reduced and the effluents from fishponds, established in the drained areas, caused eutrophication. The diversity of the flora and fauna in the lake itself was greatly reduced, but it was hoped that some elements survived as isolated pockets in the springs. Due to shrinkage of the peat and frequent flooding, the gains for agriculture in the reclaimed land decreased. Plans are now in hand to install settling ponds on the outlets of fishponds to improve the water quality, and it is proposed to construct a complex of recreational water bodies in the old marsh area, from which better quality water, of spring origin, will enter the reserve. It is also expected that a certain proportion of the original flora and fauna will establish in the recreational water bodies (Dimentman *et al.*, 1992). Lake Hula has been subjected to heavy anthropogenic pressures and indicates how, through management and redirection of the land use, partial restoration of its biological interest might be achieved. Unfortunately, many interesting endemic species have been lost.

6.3.2 Flood plains

Management of wetlands for nature conservation has often led to conflicts with local peoples, who may have traditionally used them for other purposes, such as grazing of domestic livestock. Recent thinking has moved towards community involvement in the planning of the integrated land-use of such reserves so that local interests can be incorporated in the development plan and endemic people may understand the benefits they may obtain from wise management. In this way moneys that may be obtained from the reserve management can be used to provide services for local communities and their people will not feel aggrieved that they have been deprived of a traditional resource. Excellent examples of this are the pilot studies that are being carried out on the Kafue Flats and in the Bangweulu Basin, Zambia (Jeffery *et al.*, 1992). These are both floodplain grasslands, with areas of permanent marshland, the former covering 6500 km^2 and the latter 11 900 km^2, which are inundated to a depth of 1–3 m during the wet season. They harbour spectacular numbers of wild ungulates and birds, Bangweulu being renowned for the black lechwe, *Kobus leche smithemani*, and sitatunga, *Tragelaphus spekei*, and Kafue for the Kafue lechwe, *Kobus leche kafuensis*. The clay soils are difficult to cultivate and the main food resources for the local human populations are wild ungulates, cattle grazing and fishing. Reed species are gathered for thatching and making baskets and the roots of these and water lilies are eaten. A task force has been established with representatives of the wetland project, interested government departments, chiefs of villages, and so on, to elaborate a detailed integrated land-use plan. For Kafue this involves the controlled harvesting of Kafue lechwe and zebra, which are the principal game animals, to provide meat for nearby townships and villages. In 1964, 487 lechwe were taken, which is a considerable contribution to local meat supplies. Surplus meat is dried to make biltong. Limited cattle grazing is allowed on the flats during the dry season and traditional activities, such as reed cutting and fishing for the ten species of fish that are of commercial value, continue. In this way, the vegetation of the floodplains is not degraded and the populations of wild animals are sustained.

Figure 6.7 A raised walkway to allow controlled public access in Tablas de Daimiel nature reserve, Spain.

6.3.3 Peatlands

Peatlands grow at an extremely slow rate because of their poverty in nutrients and their severe climatic conditions. They have frequently taken over 10 000 years to reach their present state of development, passing through various climatic cycles. It is therefore extremely difficult to reconstitute a damaged site, as the process of repair may well extend beyond the lifetime of the manager, or the climatic conditions for growth, particularly amounts of rainfall, may no longer exist.

The key to the management of a peatland is to maintain the hydrological conditions. Except in ombrotrophic conditions this implies control of the catchment area and, in all situations, it is advisable to have a buffer zone surrounding the bog. It is imperative to prevent an increase in the nutrient level of water entering it and to prevent windborne fertilizers or pesticides from drifting in from neighbouring agricultural or forestry ground.

Where the public have access to valuable peatlands and marshes it is advisable to construct wooden walkways to prevent damage by trampling (Figure 6.7).

For soligenous mires, changes, particularly reductions in the quantities of inflowing water, can have negative effects, possibly leading to colonization by trees and other floristic changes. Most peatlands should be protected from grazing and trampling from domestic livestock, which can suppress some plant species by selective grazing or by trampling and breaking the fragile surface. Soligenous bogs are less susceptible and can generally withstand light grazing pressure. Fire can cause damage to the vegetation and, if it penetrates into the underlying peat, irreparable damage can be done. The principal actions of management of bogs are, therefore, to set up an efficient buffer zone, either by purchase or agreement, and to protect the site from the impacts outlined above.

Much peatland management has been associated with the development of peat bogs that have already been heavily exploited by man, usually for peat extraction, and which have subsequently become flooded to form shallow lakes. These are generally managed for fishing, boating or nature reserves or a combination of these. The management as nature reserves is similar to that for gravel pits (section 6.3.4) and the Norfolk Broads, mediaeval peat cuttings which have evolved naturally, as shallow lakes and wetlands over several centuries, are good examples of the potential value of these sites as nature reserves.

Björk and Digerfeldt (1991) investigated the existing impacts and proposals for the use of two Jamaican peatlands, the Black River and Negril Morasses. Estimates had been made that the use of their peat, over a period of 30 years, could save 30% or more of Jamaica's fuel consumption. Historical and palaeontological studies were carried out to show the long-term evolution of the bogs and the more recent influence of man's agricultural and forestry activities. The Black River Morass was less affected by man than the Negril. The upper part of the river basin has developed into peat bog under freshwater conditions, with *Cladium* and more recently *Typha* dominating, whereas in the lower basin mangrove swamp has predominated, in proximity with the sea, with intermediate conditions between the two. As a result of their studies Björk and Digerfeldt proposed that the mangrove swamps, and the most viable areas of *Cladium* and *Typha* peatlands, should be conserved. Peat extraction should take place in more degraded areas of the latter, so as to produce a series of lakes where the soil types of the new bottom would be suitable for the development of aquatic vegetation. These should have irregular shores and islands. As there will be a gradient from fresh to brackish water in relation to distance from the sea, different biotopes will be formed. Freshwater flow through some of the new lakes can be manipulated *via* intakes from the rivers. A combination of preventative and rehabilitation measures will here produce a mosaic of very diverse wetland habitats, which have the potential to be of high conservation value. The authors propose that after redevelopment both areas should be made into national parks.

The value of this exercise is that the planning of the rehabilitation was done before the exploitation of peat, when the greatest number of options were open. The form of the new water bodies is then constructed as exploitation goes along. After exploitation, the options are reduced, and are often very limited as heavy excavating machinery is no longer there and the cost of reassembling it is high.

6.3.4 Artificial waters

The natural flooding of mediaeval peat cuttings in East Anglia, England, has created large areas of shallow water known as the Norfolk Broads, which have become of high natural history value. They are subject to heavy tourist pressure from boating and to eutrophication due to agricultural runoff. The problems and the integrated land management that has been carried out to overcome them are described by George (1992).

The excavation of sand and gravel from river plains for construction purposes generally results in the creation of shallow water bodies as the extraction descends below the water table. For economic reasons the excavation usually goes down to the maximum depth of the sand or gravel, leaving a flat bottom and steep banks (Figure 6.8) to the gravel pit, with little or no littoral.

Physically, this is of limited interest for the development of a diverse aquatic flora and fauna, but with suitable management it can be modified to provide the necessary conditions, particularly as water quality is often reasonably good as it is filtered by the sands of the flood plain. Recommendations for the design of such areas to promote their value for wildlife are given in the report of the Technical Working Party on the planning of an aquatic nature reserve in the Cotswold Water Park (Morgan, 1975). If possible, it should be arranged for the required modifications to be carried out by the gravel company either during or at the end of exploitation, depending on their method of working. The shoreline should be bulldozed to create a series of sheltered bays and more exposed headlands (Figure 6.9).

This should be done in such a way as to leave a maximum slope of 1:10 of the substrate from the shore to deeper water. If possible, the slope should be more gradual in places to create feeding habitat for wading birds and to allow establishment of emergent vegetation in sheltered places. Depending on the aims of management the shoreline can be managed further; for example a shingle shore may be constructed for ground-nesting birds such as plovers, or a sand cliff where bee-eaters, kingfishers or sand martins can make their nesting holes (Figure 6.10).

If one of the aims is to encourage water birds it will be necessary to construct islands to provide resting and breeding sites. These should be made in the rough form of a cross, or two crescents back to back, to give shelter from the wind from whatever quarter it might come. The islands should not be too high and their shorelines should slope gently to allow easy access. They should be planted with wetland trees and shrubs to stabilize

Figure 6.8 The vertical banks of a flooded gravel pit, with little or no littoral zone.

them and provide nesting cover. A covering of topsoil on the islands will encourage the establishment of plants. Such islands are required as resting areas by dabbling ducks, herons and cormorants. Some islands should be kept free of plants as 'loafing' areas for aquatic birds and nesting areas for plovers and terns by laying heavy-duty polythene, of the type resistant to ultraviolet light, below the surface. Ideally, this should be covered with sand expelled from the gravel-washing plant and/or gravel. Musil (1973), drawing on experience gained from Czechoslovakian fishponds, found that islands with a maximum area of 0.3 ha were optimal for waterfowl. Where it is not possible to construct bird islands a second-best is to provide rafts (section 9.3.3), but these are less aesthetic.

In order to attract *Gallinago, Tringa* spp. and other waders, Harrison (1982) carried out modifications to the land around an old gravel pit. A

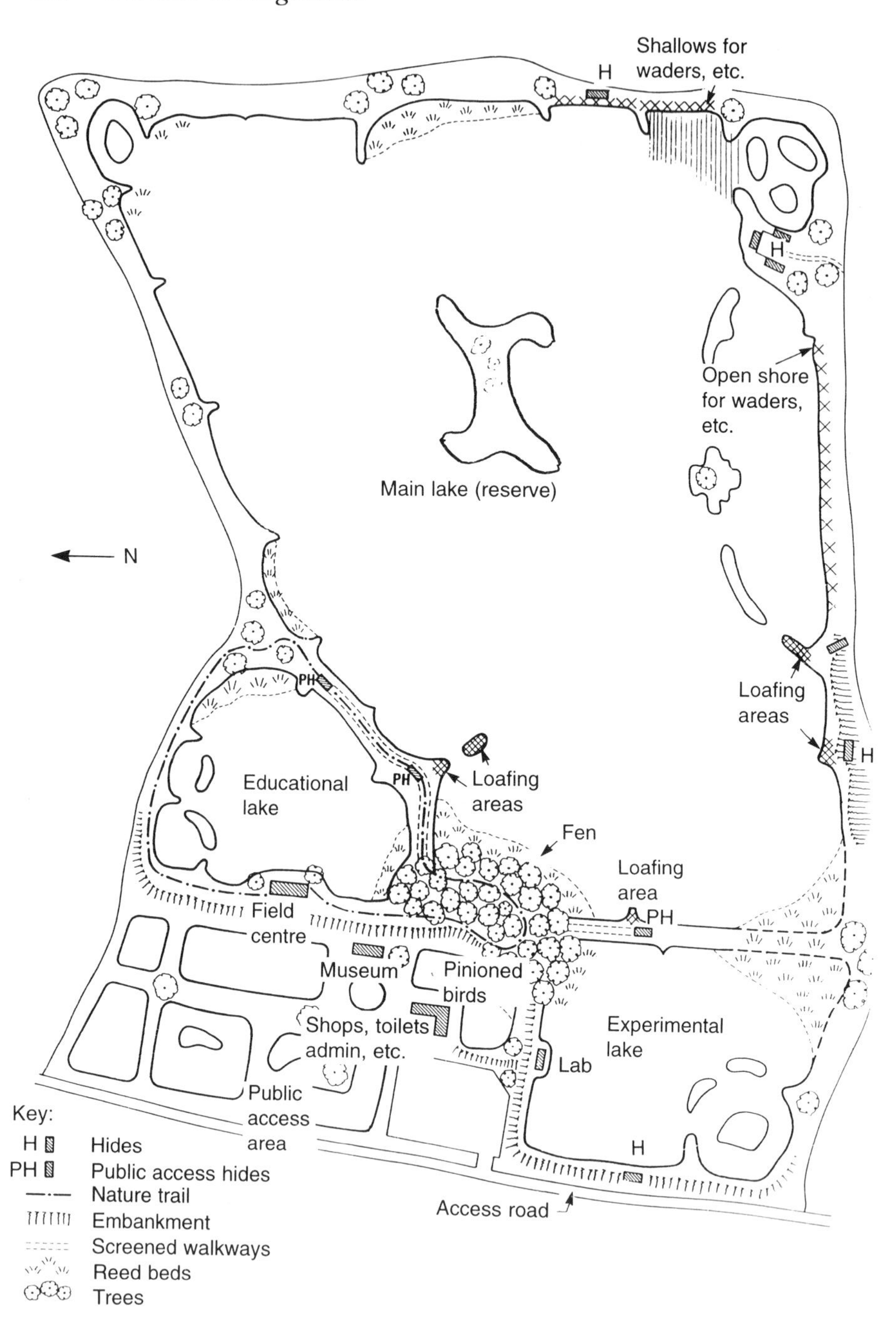

Figure 6.9 A plan for the modification of a gravel pit in the Cotswold Water Park that incorporates a wide variety of features for both wildlife and the public.

Figure 6.10 An ancient gravel pit at Lechlade, England, managed to produce a site of ecological interest.

series of shallow pools and islands was created using a string of explosive charges of 0.2 kg gelignite placed 0.33 m deep in the sand 0.66 m apart. This created about 2000 m^2 of wetland, over which 30 000 kg of cattle slurry was spread to encourage the development of a rich invertebrate fauna. Subsequently, a flame-thrower and hand weeding were used to prevent tall plant cover developing. Where this area touched the lake shore wooden booms consisting of tree trunks were staked into the sand to make other pools. A neighbouring wet area was grazed by cattle to keep the surface soft and broken and was the main feeding ground for *Gallinago*. The wader use of this gravel pit increased by 50% as a result of this treatment. If the water level can be controlled with sluices, periodic lowering of the water level, particularly in winter, is advantageous to waders, giving them access to normally submerged food supplies.

Aquatic vegetation can be established fairly easily in the open water, provided the species are adapted to the trophic conditions and substrate. Submerged species will usually propagate from stems thrown into the water but emergent species are best planted and fixed in position by staking, or protecting with booms until established. If management is biased towards aquatic birds, plant species that provide seeds for granivores should be planted.

Invertebrate species with winged adults and other means of dispersal will arrive and establish themselves in the habitats created. Corixids and chironomids can appear within a few days of providing suitable conditions. Others will have been introduced inadvertently with plant species. If particular less mobile species such as certain macrocrustaceans or molluscs are desired, deliberate introductions can be made from other water bodies. If there is no access to other water bodies fish species will usually have to be introduced but, under certain circumstances, alevins can pass through gravel beds from a considerable distance. Amphibia will obviously colonize by themselves if conditions are suitable (section 9.5).

Water bodies created in this way can be of considerable conservation interest depending on their location: for example, they are especially valuable in an area with few natural water bodies, or on a migration route for aquatic birds. They lend themselves to educational purposes, particularly if combined with an exhibition centre on the theme of wetland values and a museum (Figure 6.9). The latter can easily incorporate live exhibits in aquaria of plants, invertebrates and fish found in the lake. A useful addition is a small lecture theatre where talks can be given to schoolchildren. It may be possible to have a corner of the lake where they can catch the fauna under supervision. A dyke can be built to screen the reception area from the lake, with hides incorporated in it for viewing waterfowl, and screened walkways can lead to other hides. Captive waterfowl have been included in some educational sites to provide close-up views to familiarize the public with species likely to visit the sites. They also serve to decoy wild birds into the area where they can be most easily observed (Harrison, 1974).

More sophisticated systems can be constructed to illustrate particular habitats, such as in the Bavarian Woods Nature Reserve (Figure 6.11), where an artificial peat bog has been created in an impermeable depression by piping in acid water from some distance away, providing a substrate of peat and planting appropriate bog plants. Peatlands are, however, much more difficult to create than marshlands, because of the relative slowness of development, and require considerable finesse.

Figure 6.11 Artificial peat bog constructed in the Bayerischer Wald nature reserve.

7

Lake management

7.1 INTRODUCTION

Standing waters exhibit great variety, ranging from small shallow temporary pools through ponds and lakes to enormous waters that may be over 20 000 km^2 in area and 500 m in depth (Figures 1.3, 1.5, 7.1–7.6).

Such large water bodies, often straddling two or more countries, pose significant management problems, which can often only be solved by international agreements among politicians. Superimposed on the size and shape of a water basin are important regional differences, especially those related to geochemistry and climate. Depth is one of the most important characteristics of a lake, because on it depends the proportion of the lake's volume that receives solar radiation. The heat provided by this determines any thermal stratification and stability of the lake. Solar energy is also used in photosynthesis, which forms the basis for the productivity in standing waters. Because most energy is absorbed within the uppermost 3 m of water, shallow waters absorb almost as much solar radiation as deep ones.

In catchments where precipitation is greater than evaporation, standing waters have an outlet from which water eventually finds its way to the sea. Water in such basins is constantly renewed; consequently nutrient salts do not accumulate and the water stays fresh. In catchments where evaporation is at times greater than precipitation (as in all areas of inland drainage), higher lakes are flushed periodically but lower ones are not; the latter accumulate dissolved chemicals and are commonly called salt lakes. These are not within the scope of the present book.

In their individuality, lakes may be compared to oceanic islands (Murray, 1910); just as an island presents peculiarities in its rocks, soil, fauna and flora due to isolation by the ocean, so do lakes have individuality and peculiarities in physical, chemical and biological features due to their position relative to catchment drainage and their separation from other standing

Figure 7.1 A frozen arctic lake in Finland.

Figure 7.2 An example of a high mountain lake: Lake Isabelle, at an altitude of 3350 m in the Rocky Mountains of the USA, is covered by ice for 8 months of the year.

Figure 7.3 A temporary tropical pool in Botswana, which is of great importance to the local fauna for drinking and bathing.

Figure 7.4 Sebhet Djendli, Algeria, which is seasonally flooded each year.

Figure 7.5 Chott Fedjadj, Tunisia, is flooded irregularly, often after long dry periods during which salt deposits are produced.

Figure 7.6 Loch Fiart, Lismore, is a clear marl lake, unusual in Scotland.

waters by land. The endemic species of invertebrates and fish found in very old lakes (e.g. Lakes Malawi and Baikal) result from such isolation. Thus, in considering lake management, each water must be viewed individually and the strategy for its management must be based on scientific knowledge. In addition, many lakes are subject to a range of human pressures and though each may not have a major impact in itself the cumulative effect can be disastrous (Bauman *et al.*, 1974).

7.2 LAKE CLASSIFICATION

Various schemes for the classification of standing waters have been suggested and, though even the best of these is open to criticism, most of the systems so far presented have been more successful than similar schemes for running waters (section 8.3). This is mainly because each body of standing water is an entity with characteristic and reasonably uniform physicochemical and biological components. Running waters, on the other hand, can rarely be precisely defined and often exhibit a wide range of conditions and communities within a single system. The classification schemes proposed for standing waters so far have been based on a variety of parameters, including type of origin, physical (especially thermal), chemical and biological characteristics.

Classification according to origin has been discussed; three main types of basin are distinguished: rock, barrier and organic. Within each of these are found further subdivisions. Though this classification is a useful one and extremely practical in that most lakes fall distinctly into one class or another, it has two main disadvantages. Firstly, many lakes have changed dramatically since they were first formed, and are still doing so. This difficulty is solved to some extent in the classification suggested by Pearsall (1921) by arranging lakes in an evolutionary sequence. Secondly, lakes with entirely different origins can be similar from an ecological point of view and *vice versa*.

Lakes can be arranged according to their superficial areas, the volumes of water they contain, their mean or sometimes maximum depths, their latitude, altitude and salinity. All these classifications must be regarded as more or less artificial and, while useful in comparing a large number of lakes, are not of great value from the ecological point of view, mainly because each lake type grades into the next without discrete divisions. The points at which divisions are made for descriptive purposes are, therefore, arbitrary ones.

One of the most useful biological classifications of standing waters was originally suggested by Thienemann (1925) and later elaborated by others. The scheme suggests three major types of open water – oligotrophic (Figure 7.7), eutrophic (Figure 7.8) and dystrophic (Figure 7.9) and some of the major differences between them are listed in Table 7.1.

Figure 7.7 A clear-water lake: Lake Pukaki on the South Island of New Zealand.

Figure 7.8 A shallow equatorial eutrophic lake: Lake George, Uganda.

Figure 7.9 Loch Tallant, a typical dystrophic lochan in Scotland. Peaty waters like this are found in acid areas in many parts of the world.

Oligotrophic lakes are nutrient-poor, usually deep, clear lakes that never have oxygen deficiency; the chironomid midge *Tanytarsus* is often dominant and the culicid midge *Chaoborus* absent. Eutrophic lakes are nutrient-rich, usually shallow, turbid lakes that may have an oxygen deficiency in deeper water at some times of the year. The chironomid *Chironomus* is normally dominant and *Chaoborus* is present. Dystrophic lakes have variable amounts of nutrients but high amounts of humus, making the water brown; they are usually shallow or only moderately deep, when they may show oxygen deficiencies in deeper water. Mesotrophic lakes are intermediate in character between oligotrophic and eutrophic ones. Though this classification does not really cover all types of standing waters (Figure 7.10) and is extremely general and in some ways arbitrary, it has proved its value over a long period and the typology described is widely used.

No system of classification is ideal. Nevertheless, many are of value and, if the systems themselves are defined and understood correctly, they provide useful methods of categorizing and comparing lakes. It is probable that some combination of the systems discussed above is the most acceptable way of defining a lake quickly; for example some lakes can be edaphically oligotrophic but others only morphologically so. To describe two lakes as eutrophic may be insufficient when one is rather deep and dimictic, with a heavy precipitation of marl, while the other is shallow and oligomictic. In spite of complications, however, the schemes of classification for standing

Table 7.1 Generalized characters of oligotrophic, eutrophic and dystrophic lakes (after Maitland, 1990)

Character	*Oligotrophic*	*Eutrophic*	*Dystrophic*
Basin shape	Narrow and deep	Broad and shallow	Small and shallow
Lake substrate	Stones and inorganic silt	Organic silt	Peaty silt
Lake shoreline	Stony	Weedy	Peaty or stony
Water transparency	High	Low	Low
Water colour	Green or blue	Yellow or green	Brown
Dissolved solids	Low, poor in N	High, much N and Ca	Low, poor in Ca
Suspended solids	Low	High	Low
Oxygen	High	Low under ice or thermocline	High
Phytoplankton	Many species, low numbers	Few species, high numbers	Few species, low numbers
Macrophytes	Few species, rarely abundant	Many species, some abundant	Few species, some abundant
Zooplankton	Many species, low numbers	Few species, high numbers	Few species, low numbers
Zoobenthos	Many species, low numbers	Many species, high numbers	Few species, low numbers
Fish	Few species	Many species	Very few species, often none

waters are more successful than those for running waters and are thus more useful in a management context and essential in making comparisons to evaluate the conservation status of lakes (section 4.2). The range of the classification will depend on the scope of the evaluation – that for a relatively small geographic area will normally be more restricted than that for a large area, where variation in lake types will be greater. Valid assessments can only be made between lakes of similar type and representative lakes from each type should be chosen to obtain the full range of biological variation present. It would not be valid, for instance, to evaluate a eutrophic lake against an oligotrophic one, but representatives of each type should be included in a complete reserve series.

7.3 MANAGEMENT TECHNIQUES

7.3.1 Water regime

(a) Turnover period

Changes in the turnover period of water in a lake can have considerable consequences for its ecology and, where it can be controlled, this can be

Figure 7.10 A billabong in Australia: these small water bodies are vital to much of the local wildlife.

used for management purposes, e.g. in lakes that have become very eutrophic. Because nutrient uptake by algae is approximately proportional to nutrient concentration in the water, the dilution of eutrophic waters with nutrient-poor waters will reduce potential primary productivity and subsequent algal blooms (Henderson-Sellars and Markland, 1987). The nutrient-poor water added to achieve this also reduces the residence time of the nutrients in the lake and increases the washout of phytoplankton. Under certain circumstances, the decrease of turnover time will enhance periphyton and macrophyte growth, which themselves will help to reduce nutrients.

Restoration by dilution was implemented successfully at Moses Lake in Washington, USA which had been artificially enriched by irrigation runoff water (Welch and Weiher, 1987). The input of nutrient-poor river water caused the phosphorus concentration in the lake to drop from 200 mg/m^3 to 30 mg/m^3 and the chlorophyll-*a* concentration from 100 mg/m^3 to 10 mg/m^3. Similarly, at Green Lake in Seattle, water treated to reduce nutrients added to the originally eutrophic lake altered the residence time and resulted in a rapid change to a mesotrophic lake. The phosphorus concentration dropped from 60 mg/m^3 to 20 mg/m^3 and the chlorophyll-*a* from 50 mg/m^3 to 10 mg/m^3.

(b) Water levels

Variation in water level and its seasonal pattern can have a profound effect on the ecology of the littoral zone of a lake (Baxter and Glaude, 1980;

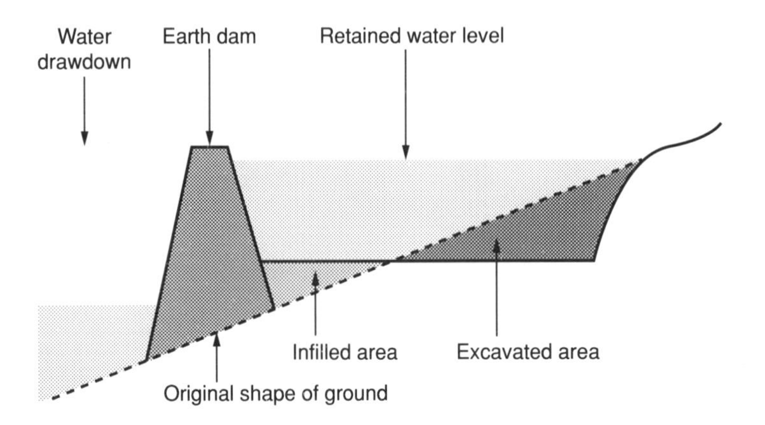

Figure 7.11 Improvement of reservoir littoral zones by bunding (after Morgan, 1972).

Nilsson, 1981) and, unlike many other important criteria, water level is one factor that can often be controlled with some accuracy. Thus, where this is possible, the management decision as to which regime is required to produce the ecological conditions desired remains to be made.

Smith *et al.* (1987) studied a group of lakes and reservoirs with widely differing water level regimes. The results were clear: in natural lakes and those reservoirs with minor fluctuations in level the littoral macrophytes and zoobenthos were varied and abundant but in waters with major fluctuations in water level the flora and fauna were very impoverished and sometimes completely absent. The study concluded that, in the types of temperate lakes studied, the permissible limits of water level fluctuation approximate to maxima of 0.5 m weekly and 5.0 m annually. Fluctuations greater than these are likely to degrade the littoral biota significantly.

Where water levels cannot be controlled within these limits, a variety of suggestions have been put forward to ameliorate the worst effects on the biota. In the main, these have proposed either bunds or depressions in the littoral zone (Morgan, 1972), which would hold back some water at lowest drawdown levels (Figure 7.11). These would be left as isolated pools (and therefore havens for many types of flora and fauna) until the water level rose again.

7.3.2 Littoral zone

The shores of many lakes are a major attraction for humans and a wide range of recreation can take place there. Increasing leisure time and mobility in different parts of the world has led to many easily accessible lake shores being used very intensively. Some shores have a naturally high resilience and can absorb considerable pressures from visitor use. However, many shores have a low resilience and can degrade rapidly (Figure 3.8). It is ironic that the basic quality that attracted people to such shores in the first instance is damaged by their overuse. Fortunately, such deterioration can often be halted and reversed by site management if the process is recognized in time (CCS, 1986).

The management of lake shores falls into three different categories.

Firstly, if the shore is a sensitive one and important in conservation terms, the recreation pressure may be diverted from vulnerable areas to areas with a high resistance and lower conservation value. Thus visitor management can be used to alleviate many of the problems of overuse of a shore.

Secondly, at some sites, an engineered bank protection may be feasible. Depending on local circumstances, and the need to maintain a natural appearance, the materials used in bank protection can range from local stone and gravels to geotextiles covered by soil and gravels. In some situations, vegetated gabion mattresses can be used on sloping banks.

Thirdly, although the diversion of public pressure may allow some damaged vegetation to recover naturally, the opportunity can be taken to replant in order to speed up the process and conserve the existing landscape. Some sheltered lake shores with a silty substrate would naturally have a reed fringe, which helps to bind the shore substrate, but this is rapidly destroyed by trampling by people or cattle. Such damage can be repaired by replacing the vegetation with clumps of the appropriate species from another site. Temporary protection of the plantings may be necessary during the establishment phase, to avoid washout by waves or heavy grazing by waterfowl. This can be accomplished by means of appropriate fencing or netting.

The shorelines of many of the world's lakes have now been significantly altered by engineering developments of all kinds – hydroelectric dams with excessive drawdowns (Figure 7.12), road foundations and embankments, jetties, marinas, concrete promenades, etc. Even the large alpine lakes, such as Lago Maggiore, now have more concrete than natural substrates along their shores (Grimaldi and Numann, 1972). Such developments, as well as being unsightly, have important effects on the littoral zone, destroying many habitats, including important fish spawning areas (Figure 7.12). In some lakes (e.g. Laghi Briantei) over 60% of the shoreline is now artificial.

Much of this kind of damage can be prevented if ecologists are involved at the planning stage so that potential problems can be explained to the engineers involved. One of the major threats is from road building alongside lake shores and here engineers frequently choose the cheapest option of road construction, often cutting into the lake shore and commonly crossing bays by infilling. At present, however, with the 'green' image now regarded by many firms as of commercial importance, many planners and engineers

Figure 7.12 The exposed barren shore of a hydroelectric lake after drawdown: Loch Errochty, Scotland.

are willing to discuss possible options that are less harmful to the lake concerned, or at least to ameliorate damage caused by replacing, for example, spawning substrates where these have been destroyed.

7.3.3 Water quality

(a) Sewage disposal

Though there are several ways of dealing with sewage, the most common system involves transport in a hydraulic system, after which it may be discharged directly to a water course, or to the sea, or subsequently treated and then discharged. In some sewer systems, liquid wastes and drainage water (from roads and rooftops) are combined, in others they are partially or completely separate. The latter system, though more expensive, is more efficient because smaller volumes of sewage need to be treated and there is no overflow into lakes and rivers during storms.

In this pollution-conscious era there is increasing pressure to treat sewage before discharge. The standard of effluent required depends on the capacity of the receiving water for self-purification and on the downstream use of water. Modern sewage systems and sewage treatment plants are expensive and the capital cost of their installation is the main factor against their development. Before the installation of new sewage plants, careful analyses should establish the quantity and quality of materials requiring treatment, and how these may vary in time. With this information it is then possible to decide on the type of treatment plant needed and the time required for its various processes, which may be considered in four stages: preliminary treatment, sedimentation, organic breakdown and final effluent treatment, followed by the disposal of the purified effluent.

Most sewage effluents are discharged directly into natural waters. Formerly it was rare for there to be restrictions on disposal into the sea, and so most of the sewage discharged in this way underwent little treatment. However, obvious pollution of the sea is now forcing authorities to consider treatment. Most inland discharges are to rivers and in many countries their quality is controlled by law to ensure that at a minimum the water course remains aerobic and large sludge banks do not develop.

Though the wastes from many industries (e.g. dairies, abattoirs, canneries, tanneries and breweries) are troublesome because of their high organic content, they can be mixed with domestic sewage and treated conventionally. Other industries, such as mining and sand-quarrying, may produce effluents with large quantities of suspended inorganic solids which can be removed by special settling tanks. The most difficult wastes are those containing toxins (e.g. from paper-making, gas liquor and chemical trades); for obvious reasons these cannot be discharged into normal treatment systems which depend on biological activity, and they must undergo special chemical treatment to remove the toxins.

(b) Eutrophication

Although the increasing fertility of lakes as they age is a natural phenomenon this process normally takes centuries. However, the phenomenon of accelerated or 'cultural' eutrophication can take place very rapidly if phosphorus and nitrogen are washed into the lake from human activities in the catchment (Hasler, 1947). The process of eutrophication has been described in section 3.1.3(b) and it is one of the most important management problems facing conservationists today. However, extensive research has been carried out on ways of avoiding or ameliorating the problem and a number of successful techniques are now available.

Henderson-Sellars and Markland (1987), in their review of lake eutrophication, outline three main methods and a number of minor ones that are available to reduce, stop or reverse the eutrophication process.

The reduction of nutrient loading by source control is probably the most realistic and successful method of controlling eutrophication. It can be implemented in a variety of ways and in some countries (e.g. Sweden and the United States) is enforced by law. Nutrient reduction can be accomplished in a variety of ways: by treating the water to remove nutrients before it reaches the lake; by altering agricultural land use (e.g. banning phosphate fertilizers) and domestic practices (e.g. banning phosphate in detergents), by diverting nutrient-rich waters away from the lake, etc. As an example, Lake Washington in the United States became very eutrophic, mainly as a result of discharges from ten sewage treatment plants in its catchment. The diversion of all of these discharges directly to the sea allowed a rapid recovery of the lake to its original oligotrophic status (Edmondson, 1991).

The removal of nutrients from the water by flushing to increase the turnover time has been discussed above (section 7.3.1(a)). In addition, it has been common practice to remove sediments, which normally contain much of the phosphorus present in a lake. This is done by extracting the top layers of sediment, which are the most recent and therefore the richest. Dredging may be carried out hydraulically, by sucking up surface sediments, or by bucket dredging. Lake Trummen in Sweden received sewage and industrial discharges for about 30 years and changed rapidly from an oligotrophic to a eutrophic system (Björk, 1972; Andersson *et al.*, 1975). Significant changes in its fish community took place during this period. The extensive layers of rich sediment that were deposited were so great that, although the sewage was eventually diverted, the lake showed no sign of recovery during the following decade. Because of this, the rich surface sediments (amounting to some 300 000 m^3) were suction dredged in 1970 and 1971. Following this, the concentration of nutrients decreased considerably and oxygen conditions improved. Blooms of blue-green algae disappeared and transparency increased in summer. The sediment removed was used to improve the nutrient-poor soils of the area and fish communities have recovered so that sport fishing is again important.

The goal of making the nutrients present in the lake water less available for photosynthesis has been accomplished by adding chemicals to precipitate them or to convert them to a less biologically available form. Copper sulphate has been used to remove algal blooms but has harmful side-effects. Dosing with alum is more successful, the alum floc removing material as it sinks to the sediments and the effect lasting for 1–2 years. This technique has been used successfully at West and East Twin Lakes in Ohio (Cooke and Martin, 1989).

Other methods of tackling eutrophication include the harvesting of algae and macrophytes, biological controls using fish (e.g. Chinese grass carp, *Ctenopharyngodon idella*), aeration of the hypolimnion, removal of hypolimnetic water and sealing the lake sediments (Henderson-Sellars and Markland, 1987).

(c) Acidification

As noted in section 3.1.3(c), many fresh waters in Scandinavia, North America and the British Isles have lost their fish populations over the last three decades because of acidification, and altogether many thousands of individual stocks have disappeared (Haines, 1981; Maitland *et al.*, 1987).

Various ways of ameliorating the impact of acid precipitation have been investigated, most of them involving the addition of calcium in some form, either directly to the water body or to the catchment of the system involved. Most of the pioneering work in this form of habitat restoration has been carried out in Scandinavia.

In Great Britain, various attempts at liming to ameliorate freshwater acidification have been attempted, most notable among which has been the work at Loch Dee (Burns *et al.*, 1984) and later at Loch Fleet (Brown *et al.*, 1988), where the former healthy population of brown trout started to decline during the 1950s and became extinct during the 1970s (Maitland *et al.*, 1987). In 1984, at Loch Fleet, a restoration project costing over £1.5 million was initiated and calcium carbonate was added to the catchment in various ways. The loch responded quickly and the pH rose from about 4.5 to 6.5 within a few weeks; at the same time the amounts of aluminium in the water decreased. Adult fish were introduced to the system in 1986 and these subsequently spawned successfully.

This experiment verified the earlier work of others, but though successful it is temporary (it is likely that further lime will have to be added to the catchment by the end of the century) and very expensive. In addition, liming often creates other ecological problems (e.g. damage to acidophilous vegetation) and it is by no means a complete solution to acidification (Woodin and Skiba, 1990). Thus, though providing a possible short-term answer to the acidification problems affecting important local stocks of fish, e.g. the Arctic charr at Loch Doon (Maitland *et al.*, 1991), it does not provide a satisfactory long-term form of habitat restoration.

A more recent and less harmful method is phosphorus fertilization (Davison *et al.*, 1995). This is cheaper and better than liming since it does not alter the nature of the water body (if soft, it remains soft) and, if proper conservation management is carried out, it does not lead to the harmful aspects of eutrophication.

7.3.4 Lake vegetation

The management of lake vegetation is normally required for two quite opposite reasons – either there is too much, with consequent problems for navigation, boating, angling, etc., or there is too little, and therefore no habitat for many invertebrates and fish nor food for birds. In most, but not all, circumstances, the growth (or lack of it) is abnormal and the manager wishes to restore the plant community to normal. That being the case, the most valid approach is to determine the causal mechanism for the problem (e.g. eutrophication) and deal with this. However, in practice, managers often adopt a pragmatic short-term approach that must be repeated every year.

Excessive growths of plants (both algae and macrophytes) are normally caused by high nutrient loadings and thus eutrophication; the methods of tackling this have been dealt with above. More direct methods usually involve mechanical clearance, the use of aquatic herbicides (section 6.2.5) or some form of biological control.

Weed cutting can be carried out manually, using scythes or other tools, or mechanically by specialized weed-cutting boats (Price, 1981). The procedure inevitably creates a degree of habitat disturbance and affects non-target organisms as well as the target plants (Murphy and Pearce, 1987). The removal of submerged plants from a shallow lake was shown by Rabe and Gibson (1984) to cause a loss of substrate, food and shelter from predators, the net result being a major change in the distribution of invertebrate species. It is also possible for fish to be trapped in the weed as it is removed and Haller *et al.* (1980) estimated that the replacement value of fish removed from a 65 ha lake during weed clearance was $410 000. Dredging to remove macrophytes involves removing some of the sediment as well and this is also a damaging procedure.

Methods of chemical control of plants (section 6.2.5) may often have an even greater impact on the lake habitat since they are, by definition, toxic to some biota; the dead plants are left to rot *in situ*, creating water quality problems; and the extent of the treatment is less precise because of the mobility of the herbicide residues. Wingfield and Beeb (1982) noted that terbutrine applied at the accepted treatment rate (0.1 mg/l) disrupted the normal diel cycle of oxygen and led to complete deoxygenation within a few days. Similar effects with other herbicides have frequently been noted (Murphy and Pearce, 1987).

Biological control normally involves herbivorous fish, usually the Chinese grass carp but sometimes other species. The grass carp has proved to be a useful control agent in many situations and, in north temperate waters at least (Stott, 1977), where it does not breed, can easily be kept under control. It is not the perfect answer, however, for it is a selective feeder, preferring the plants with soft succulent foliage to those with tough leaves. Its initial impact therefore is to create a shift in the composition of the macrophyte community. In addition, it eats some invertebrates with the weeds and digests only 10% of the latter, thus voiding large amounts of faecal material and allowing the recirculation of nutrients.

However, as noted by Wade (1995), the need to manage vegetation usually arises from the inappropriate use or management of the water concerned and the ideal solution under these circumstances is to develop a vegetation management strategy based on target nutrient loading, flow and shade characteristics for a given water in relation to the desired plant communities.

In some lakes the management problem concerns the reduction or even disappearance of desired macrophyte species and the aim is to reverse this process. Again, the ideal solution is to tackle the cause of the problem directly and often the culprit once more is eutrophication, which frequently favours phytoplankton at the expense of macrophytes. However, in other cases the problems and solutions may be quite different and appropriate *ad hoc* solutions must be sought.

7.3.5 Invertebrates

Relatively little can be done for the conservation of many aquatic invertebrate species by direct management. However, it is clear that by indirect management, largely by provision of the correct type of habitat, the quality and quantity of invertebrate populations can be manipulated. The importance of aquatic vegetation to many invertebrates has been mentioned in section 7.3.4. Some examples of management aimed at individual species are given in section 9.4.

7.3.6 Fish populations

The viability of many fish populations is closely related to angling activities and to commercial fisheries.

A substantial number of angling activities and management practices can cause damage to the conservation value of a lake. The removal of 'unwanted' species may not only eliminate native fish species but also completely alter the ecology of a water. Stocking with non-native stocks or the introduction of alien fish species may also create problems and change the ecological character of a site. Diseases and parasites can be introduced directly by anglers and their gear or indirectly along with stocked fish. The control of fish predators is a very contentious issue and further research is required here. Habitat management of various kinds is widely practised by anglers and can cause substantial damage to physicochemical and biological features at sensitive sites. Physical access, including the use of boats, may cause damage to terrestrial and aquatic vegetation as well as disturbance to many forms of wildlife.

Conservation management to prevent or minimize damage at a site will vary greatly according to its scientific importance and the degree of potential or actual damage caused by angling or angling management practices there. The solutions may range from a complete ban on angling through various types of restriction to no action at all. However, all relevant sites should have an angling management plan, agreed between anglers and conservationists, which will not only monitor angling and the fish population at each site but also provide the scientific basis for improved knowledge and management of sites in the future.

One of the major problems with fisheries of all kinds is that their management is rarely based on scientific principles or conceptions of conservation and sustainability, and the usual approach is a short-term one of maximizing yield regardless of the consequences to the fish populations concerned. As a result, many fisheries (and, consequently, local fish-based economies) have collapsed – especially those where multinational ownership of the fishery is involved (Wise, 1984; Keen, 1988).

In order to avoid such catastrophes, substantial efforts were made in the 1950s to develop quantitative theories of fisheries management and this led to the development of 'surplus yield' models (Ricker, 1954) and the foundation for 'cohort analysis' models (Beverton and Holt, 1957). For many years since then, the scientific philosophy preached has been that of 'maximum sustainable yield' (MSY) but this concept has fallen into some disrepute because of experiences in some fisheries (Larkin, 1977). Nevertheless, MSY is a convenient concept for discussing fishery management problems (Gulland, 1989) because it serves three distinct functions: a description of the status of fish stocks in relation to exploitation, a definable objective of management and a measure of the success of stock management.

Many workers have attempted to improve the models available and much more sophisticated statistical tools are now accessible (see, for example, Beddington and Rettig, 1984; Walters, 1986; Getz and Haight, 1989). However, some of the solutions being proposed are unrealistic for inland freshwater fisheries, for example regulation by taxes (Scott, 1979) or by the use of electronic TACometers on fishing vessels (Pitcher and Hart, 1982).

The problem in most fisheries is really a twofold one. Firstly, it is now known that fish populations (and fish production) can be very variable from year to year. The Beverton and Holt model assumes constant recruitment and is useful only under equilibrium conditions. Secondly, the statistics which are essential to successful management policies (e.g. the catch per unit effort: CPUE) are usually lacking. By implication, therefore, successful management of freshwater fish populations for their sustainable use requires regular statistics (Hocutt and Stauffer, 1980) on the populations of each species, in particular:

- the annual CPUE in the fishery concerned;
- the age structure in the catch and in the population;
- the growth rate of each year class.

Most fisheries should have a close season of some kind (usually during the spawning season) and the above data for the previous season can be analysed then. If the results are available for the start of the next fishing season, then catch quotas for the whole fishery can be set at limits that will allow a successful fishery to be sustained.

In conclusion, the successful management of freshwater fish populations to allow their sustainable use must rely on a sensible mixture of annual scientific information about each species and the extrapolation of this, using appropriate statistical models, into the allowable catch for the following year. There must also be equitable policies where international waters are concerned and a realization that, in the long term, the status of freshwater

Figure 7.13 The lake trout, *Salvelinus namaycush*, a major commercial species in the North American great lakes that was decimated by the arrival there of the sea lamprey, *Petromyzon marinus*.

fish populations is dependent not only on the quality of the water in which they live but also on the land use and other activities by humans in the catchment which it drains.

Non-sport and non-commercial fish species, not subject to the above pressures, may need little management provided their habitat remains healthy. Any management will relate to other pressures, such as those covered in Chapter 3. However, unforeseen disruptions, such as the entry of the sea lamprey to the Great Lakes of North America (Figures 7.13, 7.14), must always be kept in mind and the importance of adequate monitoring (to identify new events or trends) is important in this connection.

7.4 ARTIFICIAL WATERS

Contrary to some thinking, artificial waters do have a considerable role to play in conservation management. For example, because they are not natural and thus (initially at least) have no intrinsic conservation value, they are extremely useful for various experimental management purposes. The use of several reservoirs in Scotland to maintain 'safeguard' populations of important stocks of *Salvelinus* and *Coregonus*, as described in section 9.2.1, is just one example. In addition, reservoirs extend the range of lakes that are available for various activities, and it is therefore possible to redirect

Figure 7.14 Inspecting a lamprey trap on the Cheboygan River, Michigan, USA, part of a major monitoring programme for sea lampreys.

recreational pressures (e.g. angling, water-skiing, etc.) towards them and away from valuable and sensitive lake sites.

Abandoned gravel pits often provide nesting sites and/or wintering sites for aquatic birds.

7.5 CATCHMENT MANAGEMENT

The principal of catchment management is to integrate all activities practised in the catchment by overall planning, so as to maintain the quality of water entering water bodies within the catchment. It depends upon one authority or board being responsible for organization within the whole catchment rather than many piecemeal activities by different individuals or groups in the catchment.

Although integrated catchment management is an important and topical concept, the idea is not a new one and there are several long-established examples which have been very successful. The management of the catchment of Loch Katrine in Scotland (Chambers, 1983) is a little-known but long-established and successful example of sustainable catchment management. Since the inception of the Loch Katrine water supply scheme for the City of Glasgow in 1859, the West of Scotland Water Authority and its predecessors have owned and managed the entire catchment of this loch. The ground has been given over almost entirely to low-intensity sheep farming and forestry – both of which are run at a profit. In addition, there is a small sawmill within the catchment, several houses occupied by employees of the water board, a sustainable trout fishery on the loch and a summer pleasure steamer and visitor centre. Sewage from the steamer and visitor centre is piped away from the loch and treated within a reed bed system. This carefully integrated management has meant that the catchment has remained largely in a semi-natural condition over a very long period with little change in the high quality of the water in the loch and its tributaries.

It is to be hoped that this principle will be applied more generally in the future, as it is the only sure method of preventing the gradual deterioration of water bodies.

7.6 EXAMPLES OF MANAGEMENT

There are many examples of the successful conservation management of standing waters for aquatic wildlife. These range from the physical creation of new waters from a completely terrestrial situation (section 6.3.4) to complex management programmes for large water bodies, which may straddle several national boundaries and involve political as well as scientific considerations.

Worldwide, sport fishing is one of the major factors affecting the management of lakes (and, to a lesser extent, rivers) of all sizes. A wide range of fishery management practices can affect the waters concerned – often to the detriment of their nature conservation interest. These may include:

- **Removal of fish**, by
  - angling
  - trapping
  - poisoning
  - electro-fishing
  - netting
  - drainage;
- **Fish stocking**, as eggs, fry, juveniles or adults;
- **Fish introductions**, whether intentional, casual or accidental;
- **Accidental introduction** of diseases and parasites;
- **Groundbaiting**;
- **Predator control** of fish, birds and mammals
- **Habitat management** through
  - groynes and fishing jetties
  - artificial spawning areas
  - blocking access to inflows
  - removal of barriers
  - liming

  - cutting weeds
  - cutting bankside vegetation
  - construction of fishing pools
  - raising of the water level
  - blocking of outflows
  - addition of fertilizers
  - use of herbicides
  - use of grass carp
  - introduction of 'food' species;
- **Access to waters**, and thus
  - trampling
  - digging turf
  - disturbance
  - deposition of litter
  - lighting of fires
  - use of boats.

In order to assess the potential threat of any management proposals on a water, an initial analysis of the situation is advisable (Figure 7.15).

If angling is likely to be seriously detrimental to the scientific interest of the site then it should be banned. In most cases, however, this will not be necessary, but here an angling management plan should be developed that will ensure minimal impact on the water concerned. The outline requirements of an angling management plan are as follows.

- **Description of site**
  - location
  - size
  - physical character
  - chemical character
  - fish species present
  - other biological data
  - general features
- **Evaluation of the features**
- **Objectives of the fishery**
  - target fish species
  - bag limits
  - maximum no. anglers per day
  - average no. anglers per year
  - expected catch per year
- **Proposed management practices** (see above)
- **Proposed monitoring**
  - hours per angler per day
  - fish per angler per day (weight and length)
- **Revision of management plan** every year, based on monitoring data.

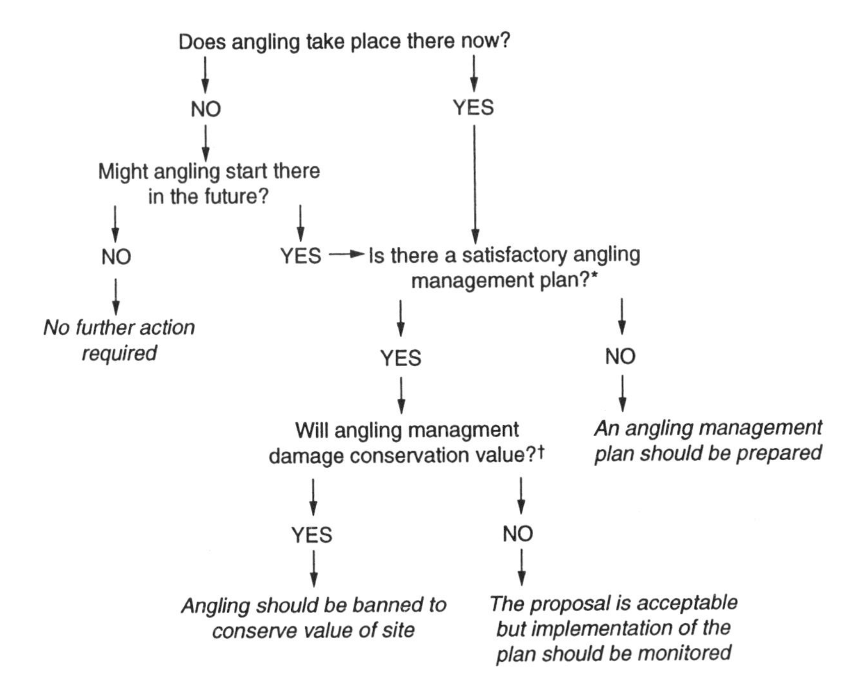

Figure 7.15 A preliminary approach to the conservation management of waters affected by angling.

7.6.1 Ponds

Experimental studies described by Giles (1992) have demonstrated the important links in the aquatic food chain among plants, invertebrates, fish and birds and how management can affect these links. Invertebrates (especially chironomid midges, both larvae and adults, and the snail *Potamopyrgus*) were shown to be a key factor for the breeding success of mallard and tufted ducks at various gravel pit nature reserves in England. However, it was also found that a critical factor in the production of these same invertebrates depended on the numbers of coarse fish at a site. Bream, *Abramis brama*, tench, *Tinca tinca,* roach, *Rutilus rutilus* and perch, *Perca fluviatilis*, were all found to have an impact on different components of the invertebrate communities and the removal of these fish from a water was followed by an increase in both invertebrates and weed beds and in the success of ducklings. The spectacular increase in weed growth after fish removal led directly to increases in the numbers of wintering mute swans, gadwall, coot and pochard.

These findings led to a series of new management practices for such waters. Since duckling survival is so critically linked to the availability of aquatic invertebrates it is essential to ensure that wildfowl breeding sanctuaries have the following characteristics freely available to broods of dabbling and diving ducks.

- Water depth should range from 50–300 cm.
- The lake bed should be covered with a good growth of submerged plants.
- Fish should be netted out on a regular cycle to maintain high invertebrate availability for ducklings.
- The key invertebrate groups to encourage are midges and snails.

Thus, if the management of a pond or lake is to promote the vegetation there then fish such as bream, *Abramis brama*, carp, *Cyprinus carpio*, and roach, *Rutilus rutilus*, should be netted out regularly, as all of these can suppress seedling establishment and growth. Conversely, if the intention is to suppress weed growth for some reason (perhaps angling or sailing) then rather than using herbicides or manual cutting, stocking with a high density of adult bream or carp (to a combined standing crop of 400–500 kg/ha) will cause a significant decline in vegetation. However, it should be noted that the resulting conditions in the pond may be turbid water, prone to algal blooms in summer (Giles, 1992)

7.6.2 Lakes

Kallemeyn *et al.* (1993) have demonstrated how the aquatic ecosystem of a natural lake affected by regulation may be restored by the use of a more natural hydrological regime. Rainy Lake, with a catchment of 38 600 km^2, is part of the Hudson River drainage system. Hydrological conditions in Rainy Lake and Namakan reservoir have been regulated by dams since the early 1900s. This regulation has removed much of the hydrological variability that the lakes would experience under natural conditions, the controlled water levels adversely affecting key elements of the aquatic system: littoral vegetation, benthic organisms, fish, aquatic birds and mammals. Changing water levels have also affected fish spawning success.

Drawing on previous scientific research (Kallemeyn, 1983) and using mathematical modelling, alternative water level regulations were developed by various user groups in the area, which were designed to imitate natural fluctuations in water levels, including annual and long-term variability. Although some components of this new programme conflicted with the needs of some water users it was found possible to develop a compromise regulatory programme that provides for both human and biological needs.

7.6.3 Large lakes

The management of large lakes is a major problem, not only because they are often bordered by more than one country but because of the enormous size of their catchments, where varying land uses may pollute runoff and ground water or where impoundment of tributaries may reduce inflow. The different riparian countries may make different demands on the freshwater resource, for water supply, irrigation, fisheries, transport or tourism; these differences can cause economic and political conflict, which impedes the development of an integrated conservation policy. The only realistic way to overcome these difficulties is to agree to set up a joint supervising commission with members from each country, supported by international bodies like FAO, WWF and the World Bank. The job of this group would be to draw up a realistic plan for the sustainable use of the lake and the conservation of all its resources.

A good example of this type of problem exists at Lake Tanganyika. This lake contains the greatest volume of fresh water in Africa (19 000 km^3) which, in a semi-arid area, is invaluable as a source of drinking water and for irrigation. It has the largest fishery for freshwater sardines *Stolothrissa tanganicae* and *Limnothrissa miodon* in Africa, producing around 19 000 tonnes per annum (Coulter, 1991), which, it is estimated, could be increased to 300 000 tonnes without harming the stock. At present, over 80% of the people living around the lake obtain their protein from this source. There have been suggestions that a large predatory fish, such as the Nile perch, *Lates niloticus*, should be introduced, which could have catastrophic results – as it has had in Lake Victoria.

The lake is an important route for boats plying between the four riparian countries, where roads are few, and it also has high scenic value.

Lake Tanganyika has the highest number of endemic species of any lake in the world, with over 250 species of fish and over 250 species of molluscs and crustaceans. There is a thriving export trade of live fish, particularly cichlids, for the aquarium hobby in Europe and the United States. As many of these species are restricted in distribution – to one rocky headland or another – there is danger of individual species being eliminated by overzealous collecting. One trader has sensibly started to breed captured fish in artificial ponds on the lake shore. The shore has also a high diversity of plants and animals, with 1300 known species, obviously related to the exceptional age of the lake – which is about 20 million years. The lake has an exceptionally long retention time of over 1000 years, which makes it extremely vulnerable to pollution: persistent residues would be extremely slow to wash out. Oil exploration along the northern shores in the 1980s caused great concern because, if exploitation was ever to take place, then oil spills would have serious consequences.

Table 7.2 Economic and biological value of Lake Tanganyika

Resource	*Conservation implications*
19 000 km^3 of pure water	Rational utilization for drinking water and irrigation
Important fishery for freshwater sardines	Sustainable management of the resource A ban on introductions of alien fish
Boat transport	Avoidance of pollution
500+ endemic species	Strong protection of this important scientific resource involving research and nature reserves
Aquarium trade	Careful collection procedures so as not to damage fish stocks
Reserves of oil	If exploitation of oil takes place it should not be allowed within the lake or close to the shore, where spills could cause irreparable damage Transport of oil by boat should be banned
Forestry and agriculture	Clear-felling and denudation of the neighbouring slopes should be stopped to prevent sediment entering the lake Pesticides and fertilizers must not be allowed to enter the lake
Industry	Discharges to the lake must be avoided
Shoreline townships	Domestic sewage should not enter the lake
Tourism	This depends on maintaining clean water, unspoilt scenery and local traditions

Other conservation problems are the deposition of excessive loads of sediment in the lake, caused by runoff following deforestation of the surrounding slopes. Cohen *et al.* (1993) found a reduction of 40–60% in the diversity of ostracods in the affected areas of the lake. In addition, there are threats from domestic and industrial pollution around the towns and pesticides from agriculture.

In view of the outstanding economic and scientific importance of this great lake, an international conference on the conservation and biodiversity of Lake Tanganyika was held at the University of Burundi in March 1993. This meeting proposed the formation of a Lake Tanganyika Commission to coordinate planning, conservation and research activities within the lake basin (Table 7.2).

The specific need for reserves was highlighted. Coulter and Mubamba (1993) proposed, in the first place, the extension of the four existing wildlife land parks on the lake shore to incorporate underwater parks. Research was proposed to determine where other parks should be developed to encompass entire endemic species populations. It is clear that the protection of discrete parks in a lake does not give complete protection to these areas because water, plants and animals can move freely in and out together with any accompanying pollution or diseases. However, if put into effect, these areas may serve to promote the value of conservation and may eventually extend to conservation laws for the whole lake. The authors point out that reserve areas could be used to promote tourism and for educational purposes.

Unfortunately, so far, none of these proposals has been put into effect because of logistical problems and the unfortunate civil strife in Burundi.

Another important African lake with a large endemic fauna is Lake Malawi, which has similar pressures to Lake Tanganyika, which militate against conservation there. Reinthal (1993) investigated the patterns of biodiversity of the rock-dwelling cichlid fishes and found that the genus was the appropriate taxonomic level for conservation purposes and that there is a significant positive relationship between the percentage occurrence of a genus in communities and the overall diversity within the genus. He proposed the extension of a Lake Malawi park into areas of the northern and central parts of the lake, which have high generic endemism. He emphasizes that long-term conservation can only be achieved by the establishment of an international park among the three riparian countries.

In North America, only two countries (the United States and Canada) border the five Great Lakes. Here, there is a successful international Great Lakes Fishery Commission, which has over recent decades done very valuable work in helping to manage these enormous stretches of water. Numerous problems have been tackled, for example the invasion of the lakes by the sea lamprey *Petromyzon marinus*, eutrophication of the lower lakes and, more recently, invasions by alien species (ruffe, *Gymnocephalus cernuus*, and zebra mussel, *Dreissena polymorpha*) from Europe.

Figure 8.1 The Victoria Falls on the River Zambesi at low water.

but also its place in a national classification of river types (Maitland, 1985b).

In the past, classification of British rivers has relied heavily on the use of plant communities. Haslam (1975) indicates how plant species are closely correlated with different river habitats. The assessment of the conservation value of rivers from their invertebrate communities has been more problematical because of the under-recording of many groups (Nature Conservancy Council, 1989). Even where good invertebrate data are available for a river (Maitland, 1966), the lack of comparable data from elsewhere has made it difficult to place the river in a national context. Following earlier proposals (Maitland, 1979), a classification of British rivers based on macroinvertebrates has now been developed (Wright *et al.*, 1984), largely for water quality assessment, but this will have an increasing conservation relevance as it is refined.

The range of running water types is large, sometimes several of them occurring, interconnected, within a single drainage system. This presents various difficulties in the conservation management of rivers, for they rarely form discrete units like lakes. The range covered within the running water series includes small trickles and seepages (often temporary in nature), ditches, larger fast-flowing streams and rivers and large slow-flowing rivers (Figures 1.1, 1.4, 4.3, 8.1–8.3).

8

River management

8.1 INTRODUCTION

It is clear that rivers are limited resources, which are subject to often conflicting demands from a wide range of users – yet basically all of these users want the same thing: reliable supplies of clean water. Rivers – and their entire catchments – therefore need to be actively managed in such a way as to balance the needs of the various users and to ensure that both the quantity and quality of the resource are maintained. For example, an important economic fisheries-related problem in this context is the conflict between the lower and the upper reaches of salmon and sea trout rivers. Here, the situation is that the proprietors of the lower reaches gain almost all the benefits of good stocks of adult fish running in the river, yet it is the upper reaches of the river (and its tributaries), with less useful fishing, where the fish spawn and the nursery areas occur. Thus there is little incentive to the upper proprietors to manage the river for native salmonids.

Thus, in looking at the difficulties encountered in developing any management plan we must also be clear about the advantages – both to the individuals who may appear to be disadvantaged and to society as a whole. Thus, for example, an individual farmer may lose income from his agricultural activities because of loss of land to riparian habitats, etc., but this could well be compensated naturally by income from improved fishing, some shooting and new recreation including, for example, bird watching.

As discussed in Chapter 4, important criteria used to identify rivers of conservation interest are diversity of habitats and species, naturalness of catchment and river corridor and representativeness of rivers of a particular type (Newbold *et al.*, 1983; Boon, 1991). Other factors that should be taken into account are rarity (of river type, species or communities), size or extent, fragility of habitats and communities, and geographical position (Maitland, 1985a). Evaluation of the conservation value of a river thus requires a good knowledge of not only its plant and animal communities,

Figure 8.2 An Icelandic river fed by glacial water.

The degree of slope is important. Mountain streams and rivers, with a rapid descent in altitude, are essentially eroding whereas the meandering rivers of the plains deposit the sediment eroded upstream. Middle reaches are intermediate in character, with successive pools and riffles.

The geology of the catchment underlying a running water has a strong influence on physical and chemical characteristics (Chapter 1). Further, the nature of the bedrock and soils affects the character of the aquatic substrate and also relates to the rate of erosion, and hence succession, of that running water. The suspended solids present in turn affect the penetration of light under water. The flow characteristics of running waters are also connected to geology, notably in the control exerted by rock and soil formations, and the relationship between ground water and surface waters. The flow pattern of running waters depends largely on the nature of this relationship. In

Figure 8.3 A typical turbid lowland river: the Murrumbidgee River in Australia.

addition to controlling the quantity of ground water, local geology also exerts a strong effect on its quality, notably in connection with the dissolved solids present. Crystalline rocks, which are highly insoluble, produce nutrient-poor waters with low plant and animal densities whereas calcareous rocks give rise to much richer floras and faunas.

Finally, the type of human land usage and activities, both within the catchment and on a river, can be a major controlling factor in its ecology, completely altering the natural characteristics discussed above. Catchment management therefore must inevitably come into any discussion of scientific river management.

8.2 SCIENTIFIC BASIS FOR RIVER MANAGEMENT

Any examination of the types of fresh water that are given conservation protection shows that both on a national and an international basis rivers have been sadly neglected – strong preference having been given in the past to standing waters in the establishment of nature reserves or other protected areas. This bias is probably because of the much greater difficulty in protecting rivers as opposed to lakes, rather than that rivers are less worthy of conservation than lakes.

One of the major reasons for the continued widespread degradation of environmental conditions in streams and rivers throughout the developed world is the inadequate input, transfer and application of ecological knowledge to decisions concerning the management of rivers and their biota (Swales and Harris, 1995). It is essential that adequate communication, in simple terms understandable to non-scientific decision makers, should be developed between research workers and politicians if strides are to be made in catchment protection.

An important aspect of river management which is sometimes ignored is the ability to measure the extent to which a river has been, or potentially could be, damaged by a particular human activity. A major problem is that, as noted above, rivers can vary substantially from one another and, even within a single river system, there can be considerable variation from source to mouth or from one tributary to another. Thus an understanding of the natural variation in the physicochemical conditions and biota of different river systems is an important aspect of any potential management programme.

8.3 RIVER CLASSIFICATION

Attempts to classify running waters have been of two kinds: firstly, the division of the system into definable zones (from source to mouth), and secondly, the separation of river systems into definable types. Each type of classification presents its own problems.

Various schemes of zonation for running waters have been devised; most of the early attempts at classification relate to just one or a few factors – for example, a single group of plants or animals. Harrison and Elsworth (1958) based their initial scheme of zonation on physical features such as substrate, while Tansley (1939) classified rivers into zones on the basis of their vegetation. Schmitz (1955) recognized various invertebrate zones in running waters, while one of the common methods of differentiating zones in rivers is the distribution of common fish species. Such schemes, originally described for European rivers by Thienemann (1912), were adapted for rivers in the British Isles by Carpenter (1928). In general, any rigid scheme of zonation for organisms in running waters, however clear-cut it may be, is rarely valid for other groups there, or even the same group in another water. The general theme of change in biotic associations from source to mouth in a running water is one of transition rather than zonation. The main value of characterizing zones is that, in a general sense, they may be useful for descriptive and comparative purposes.

There have been various attempts also to classify whole running water systems; Ricker (1934) classified on physical and chemical characteristics, and Carpenter (1927) on the type of origin of each system. Lagler (1949) suggested a very arbitrary system based on the average density and weight of invertebrates in a standard area. One of the most valuable discussions on the classification of running waters is that of Berg (1948), who points out that running waters cannot be classified like standing waters because they are not uniform entities, but are systems which change from source to mouth.

It is not possible, therefore, to undertake ecological groupings of running waters, but merely of certain reaches within them, and the unit in such a system will not be an entire running water channel, but a stretch of one within which the environment is uniform, i.e. a habitat. The problem of classifying running waters is thus one of describing these habitats and their organisms; there is still a dearth of information about these, for the fauna and flora of few running waters have been studied in detail from source to mouth. In relation to such studies, the concept of stream order hierarchy is a useful one.

Stream order, as defined by the modification by Strahler (1957) of the original scheme by Horton (1945), classifies a stream without tributaries as first order, a stream below the confluence of two first orders as second order and so on. The order of the outfall stream can be dependent on the inclusion or omission of a single first-order tributary and, since most calculations of stream order are based on maps, on the scale of map used. Thus the change of stream order on a sample of river systems examined at map scales of 1:625 000 and 1:63 360 was by an order number of 2 for 68% of the sample (Smith and Lyle, 1994).

A broad range of features characterizes a natural river (including the fact that its catchment should also be relatively natural). For example, its channel should not be canalized or ditched in any way and the channel substrate should consist of entirely natural materials like sand, gravels, stones and boulders and not concrete, brickwork or gabions. The water quality should be high and reflect the geology and soils of the natural catchment rather than any influence from human activities. The biota too should be characteristic of the geographical area concerned – there should be no introduced alien species nor any loss of native ones. Such systems have two major features – they are sustainable and they have a characteristic, usually high, biodiversity for each habitat represented.

Natural rivers have a high diversity of flora and fauna, which increases with both the size of the river and the diversity of physicochemical conditions within the catchment. Most rivers have at least two major biotic communities – one associated with the silty, lowland, meandering stretches, the other with the rocky, upland, fast-flowing stretches. These communities are each species-rich in themselves, have very few species in common and together make up a highly diverse biota for the river as a whole.

The particular parameters chosen in any classification of running waters will depend on the information available on morphometric, geological, chemical and biological conditions in the area studied. The main principle is to devise a system that divides the reaches of the running water into relatively homogeneous groups, thus allowing comparisons, within groups, of the conservation value of similar water bodies, and aids rigorous selection of sites for protection so as to cover the whole range of ecological variation exhibited.

8.4 MANAGEMENT OBJECTIVES

8.4.1 The river channel

Modification of the river channel through straightening and canalization has degraded many rivers, especially in lowland areas. Frequently, the river bed is so altered that it can only be returned to its original condition with some difficulty. Recolonization by plants and animals from other areas may still be possible but, because of simplification of the channel environment, the decrease in microhabitats leads to impoverishment. Schiemer *et al.* (1991) demonstrated a striking deleterious effect on the fry of rheophilic fishes from straightening the channel of the River Danube and protecting it with large rocks, compared with the unmodified situation with natural, sloping river banks. In adjacent stretches, only one species was found in association with modified banks whereas 12 species occurred in the natural sections. The strong interdependence of the Danube fishes on the main river channel, its tributaries, backwaters and areas only flooded at high water was shown by Schiemer and Waidbacher (1992). Most species lived their adult lives in one habitat and spawned in another, so that destruction of one or other habitat by channel modification could eliminate the species. Piping or ditching land to improve drainage increases the dangers resulting from higher water levels in wet weather and lower levels in dry weather, both a direct result of the faster runoff of water from the land.

The clearing of forests for agriculture also increases the runoff of surface water and the rate of soil erosion with subsequent silting in the waters draining such areas. Most types of cultivation lead to a regular loss of soils, which tend to be washed off and affect the ecology of waters into which they drain.

One important management objective in the move towards restoring wild rivers is the removal of artificial barriers to the migration of all fish species. Such barriers may be chemical or physical and each barrier must be treated on its merits in relation to solving the problem it creates.

Even where the upper parts of a river are completely natural there can be a major loss of diversity within the fish fauna due to the absence of a migratory component. This occurs where migratory fish such as lampreys, Atlantic salmon and eels are unable to swim upstream because of barriers – either physical or chemical, or both – somewhere in the lower reaches. The loss of diversity within the fish community can be substantial; for example, when pollution in the lower reaches of the River Clyde in Scotland was at its worst and several river barriers were present in the form of weirs, no

Figure 8.4 Solving one of the ecological problems created by dams on rivers: a fish pass on a Scottish river.

fish could live or pass through and the clean upper reaches had only eight fish species. To these should have been added another seven species that were unable to migrate upstream because of the various barriers below.

Physical barriers are variable in nature. Many are obsolescent low dams and weirs on rivers that were formerly used by mills and other industries that have long disappeared. Most of these could (and should) be removed to allow natural runs of former migratory species to be restored. However, thought must be given to such artefacts, which may be important in terms of industrial archaeology and heritage. Other barriers, such as existing dams for water supply or hydropower, must obviously be approached in a different way and here the usual solution has been a fish ladder of some kind (Figure 8.4).

Table 8.1 Summary of the potential temporal sequence of impacts on fish populations in a river where engineering works are proposed

Change	*Impact*
1. Engineering for new scheme	Short term silting and pollution *Detrimental to most fish*
2. Reduced flows	Shift from lotic to lentic conditions *Favours cyprinids over salmonids*
3. Lower water levels	Decrease in fish habitat *Reduces fish productivity* *Favours cyprinids over salmonids*
4. Reduced water quality	General lowering of quality standards at most sites *Favours cyprinids over salmonids*
5. Increased siltation	Smothering of gravels and pools *Favours cyprinids over salmonids*
6. Increased vegetation	Slowing of local currents; increased siltation *Favours cyprinids over salmonids*

In general, these have been designed for salmonid fish, which are good swimmers, and not for lampreys, shad and other migratory species, which swim less well, and they may require modification to allow these species to pass upstream. In general, present views on fish-pass design favour more natural channels that circumvent the barrier involved along one side.

There are substantial opportunities to reverse the decline in biodiversity that has taken place in many rivers by removing chemical and physical barriers (mostly in the middle and lower reaches) and thus allowing a variety of migratory species to return to their original habitats. In some rivers, the composition of the native fish communities could be increased by 100% in this way – a dramatic restoration to the diversity existing there 200 years ago.

8.4.2 River regulation

A large number of rivers around the world are now subject to some form of regulation. River regulation through the diversion and manipulation of flows for power generation, water supply, flood control, etc. is one of the most damaging influences on running waters and has been shown to have extremely harmful effects on the aquatic biota (Ward and Stanford, 1979; Lillehammer and Saltveit, 1984; Petts *et al.*, 1989; Table 8.1). However, recent studies, as well as indicating the extent of damage to riverine communities by flow regulation, have also provided a wealth of scientific data on which to base better management decisions in the future.

Instream flows provided after development for environmental reasons are sometimes called 'environmental flows' or 'compensation flows' and are intended to maintain the habitat and its biota in the affected stretch of river in a natural state. In practice, most compensation flows are provided for fishery interests – the latter, in the past at least, being the strongest lobby able to challenge engineering developments on rivers at the design stage.

Various techniques have been developed to assess the instream requirements of the aquatic biota in rivers. If past flow records are available, the historical discharge is used as a basis for providing a fixed proportion of flow. If no flow data are available, habitat analysis methods can be used, combining data on hydraulics and hydrology to determine the extent of usable habitat by transect analysis and hydraulic simulation. Alternatively, instream habitat modelling methods, which determine habitat preference curves for various important species, are used to model how various discharges affect habitat availability. All these methods have been subject to criticisms of various kinds (Armour and Taylor, 1991).

An expert panel assessment method (EPAM) was developed by Swales and Harris (1995) on the Murray–Darling River in Australia to provide a reliable and accurate method for assessing environmental flow requirements as an aid to better river management. The method depends on the utilization of the professional experience of specialists in fluvial sciences to assess the suitability of instream flows for river ecosystem processes. Two expert panels (each including a fish ecologist, an invertebrate ecologist and a fluvial geomorphologist) inspected the river under four flow conditions (10%, 30%, 50% and 80% flow percentiles) and ranked flow suitability on a scale of 1 (poor) to 5 (excellent) in relation to previously defined ecological criteria. The findings suggest that EPAM can be a useful tool in assessing the ecological suitability of instream flows in relation to river regulation and Swales and Harris (1995) conclude that the method provides an avenue whereby ecological knowledge can be incorporated into decisions concerning sustainable river management.

8.4.3 Water quality

Although there are still many very badly polluted rivers around the world there have been enormous advances in pollution control over the last few decades and a number of the worst rivers are now much cleaner. In North America and much of western Europe strict legislation controlling industrial and domestic discharges, with severe financial penalties for infringements, are gradually having effect and the battle to control point source discharges, though by no means over, is being won.

Thus, for example, the Rivers Clyde (Scotland) and Thames (England) are now far less polluted than they were 50 years ago and fish have been returning to them in increasing numbers. At their worst, both rivers were fishless in their lower reaches. Over the last two decades, many freshwater and estuarine species have returned to the lower Thames, which now supports a diverse community not unlike its original one. Rehabilitation of the River Clyde has been rather slower. However, salmonid fish are conclusive evidence of high water quality and the return of the Atlantic salmon to this river after an absence of more than 100 years is a tribute to decades of work by the local river purification board (Maitland, 1987a).

Though there have been these successful developments in the control of point source discharges through appropriate legislation and modern methods of effluent treatment, this is not the case with non-point (diffuse) sources of pollution, notably agricultural fertilizers, herbicides and pesticides and acid deposition. These are more intractable for individual river managers or authorities to deal with and must be tackled at a larger geographic scale – through catchment, national or even international programmes and agreements.

8.4.4 Riverine vegetation

Macrophyte vegetation is a major part of the biota and ecology of most river systems, providing much of the primary production, and thus food, as well as important shelter for invertebrates and fish. However, vegetation can pose problems for some river managers by obstructing flow, impeding navigation and causing difficulties for anglers. This has led to various methods of weed control, including weed cutting and the use of herbicides (Pieterse and Murphy, 1990), which can be extremely damaging for a variety of river biota.

However, Wade (1995) has pointed out that in many instances the need to manage the vegetation has arisen from the inappropriate use or management of the river as a system and cites eutrophication from sewage effluents, inappropriate river regulation and human settlements on floodplains as causes of the increased vegetation. The ideal solution under these circumstances is to develop a vegetation management strategy based on target nutrient loading, flow and shade characteristics for a given river in relation to the desired plant communities.

The control of invasive aquatic or riparian alien plant species requires a different management strategy, with a higher level of interference. The successful control depends on incorporating three key components into the management programme: knowledge of the autecology, a coordinated management programme and the prevention of further spread (de Waal *et al.*, 1995). Specific control may use chemicals, manual cutting and grazing, or some combination of these. Control must be carried out in a coordinated manner from the source of the river downstream, leaving no areas with living plants, or else highly successful invasive species will re-establish themselves from nuclei upstream.

Dead wood, from fallen trees and branches, can have a major impact on the hydrological, hydraulic, sedimentological, morphological and biological characteristics of river channels which are important for the stability and biological productivity of these channels (Gurnell *et al.*, 1995). This should be taken into account by ardent enthusiasts when clearing wood from rivers in the name of conservation. Only obstacles that are likely to cause damage or impede fish migration should be removed.

8.4.5 Invertebrates

As in lakes, relatively little can be done for aquatic invertebrate conservation in rivers by direct management. However, as with lakes, much can be done by indirect management; largely by provision of the correct type of habitat, the quality and quantity of invertebrate populations can be manipulated.

8.4.6 Fish populations

Most fish populations are influenced by human activity, often in a variety of ways, and there is a general assumption among anglers and commercial fishermen that fish stocks 'need to be managed' – usually not by any curtailment of their own activities but by influencing the community in some other way. Stocking is the standard 'answer' to problems, but many other activities are carried out in the name of fishery management – culling predators (fish, birds and mammals), altering the physical environment, adding fertilizers – even introducing alien fish species (Campbell *et al.*, 1994). Mostly, such management has no scientific basis and rarely are the results measured.

Fortunately, the sustainable management of wild fish populations is at last becoming a desirable goal for at least some fishery managers on various waters and the conservation management of fish stocks is potentially compatible with the aims of angling management. Thus the new objectives of such management emphasize the importance of both instream and riparian habitat management, the removal of obstacles to fish migration, proper monitoring of the catch and the education of anglers (Tweed Foundation, 1995). All of these activities fit in well with the sensible management of whole rivers and are to be commended.

8.5 RECREATION

As indicated in Chapter 3, human recreational activities of many kinds can affect river systems. The management problem here is to relate the possible impact of any activity to the river as a whole and to its most sensitive stretches and also to have a clear awareness of the conservation value of the river in local, national and international terms. Once again, therefore, it is essential to have adequate scientific information concerning the river before any sensible management strategy can be prepared.

Obviously, if the river is of high conservation importance it must be given special protection and this topic is dealt with in Chapters 4 and 5. Even then, however, it may not be necessary to ban all forms of recreation and those that will not damage the physicochemical character and biota of the river may be allowed. Other forms of recreation, which may cause physical damage or disturb the biota (e.g. power boating) may well be banned. In general, it is useful to have adequate knowledge of all the rivers in any geographic area, so that recreational pressures may be directed to those systems of least conservation value or where least damage will be caused.

8.6 RIVER RESTORATION

Because of the wide scale of habitat loss in the past there is considerable scope in many countries for successful restoration of important habitats and there are at present a number of schemes with the specific objective of habitat restoration. This may involve a number of restoration activities, particularly the elimination of pollution, the removal of artificial dams, weirs and other barriers, the reinstatement of spawning gravels, the restoration of meanders where rivers have been canalized, the provision of more compensation water where over-abstraction has taken place and the enhancement of riparian zones.

The Blanco River in Colorado, USA, was channelized after a flood in 1970 in an effort to protect riparian land from further flooding (Berger, 1992). This gave a wide, straight, flat-bottomed trapezoidal channel within a levee system. The engineered changes resulted in channel instability and the creation of a braided reach quite unlike that there previously. Numerous ecological changes followed including both sedimentation and bank erosion, summer temperatures were higher and the whole river froze in winter. Most fish disappeared. In restoring the river, the main objective was to stabilize it within a well-incised but natural-looking channel and this was done by modelling the new channel on the physical criteria of similar neighbouring rivers. The river now has new meanders and deep pools with stable banks and good riparian vegetation. Major improvements have taken place in the fishery.

Ox-bow lakes and other wetlands lying alongside rivers are among the most damaged of all habitats and the majority have disappeared through infilling and drainage. Although not generally important for fish species such wet habitats are of major importance to several botanical communities, to amphibians and to many birds and mammals and their restoration alongside any river system would add enormously to local biodiversity. In addition they would form an important additional buffer area in relation to flooding and the wash off of nutrients from land to river.

The opportunities afforded to the restoration of rivers to their former diversity and value owing to the availability of land which was formerly required for agriculture are enormous. However, relatively little action has been taken in this area, largely because the legislation involved has not been imaginative enough nor is the situation clear enough in the long term for farmers (and others) themselves to become involved in positive action.

However, the advantages of forward thinking management schemes for restoration to create 'wild' rivers are clear and numerous. For example, the destruction of artificial drainage systems within agricultural areas to be 'set-aside' (i.e. taken out of cultivation to reduce agricultural overproduction) in the long term would restore a more natural runoff to adjacent rivers. This, combined with the cessation of the spread of chemicals (fertilizers, herbicides and pesticides) on to these lands would greatly reduce the input of such chemicals to rivers. The actual geographical layout of set-aside ground could also be used wisely by maximizing strips alongside rivers; thus the use of 1 ha of set-aside land as a strip 1000 m long by 10 m wide alongside a river would be of much greater benefit than a 100 × 100 m square field. Agricultural ground which was formerly defended against floods but which now could be flooded is yet another value to the river which could result from the intelligent use of long-term set-aside schemes.

8.7 CATCHMENT MANAGEMENT

Most plant nutrients and other chemicals in the water come indirectly from rainfall that has fallen on the catchment and accumulated a variety of ions there. Thus human activities within the catchment can have a direct effect on river water quality by exposing soils to rainfall and by adding fertilizers and other chemicals, which may be washed into the river, often via drains and ditches. In the river, chemicals are used by aquatic plants in the presence of sunlight to build up plant tissue and form the basis of an important food chain, the higher parts of which are occupied by invertebrates and fish and birds. The latter then cycle many of the nutrients back into the system, either directly when they die in the water or indirectly when they die on land.

However, there are many problems with trying to develop an integrated approach to water management. As Howell (1994) has pointed out, although the World Conservation Strategy (IUCN *et al.*, 1991) listed as one of their priorities, that by 1995 'all high income countries will have established cross-sectoral mechanisms for integrated water management based on drainage basins and the application of an ecological approach', in Scotland, for example, the statutory framework does not meet this requirement. In fact, the complexity of the framework for managing fresh waters actually hinders conservation management in some areas, while in other areas statutory provision is completely absent.

In England and Wales, the National Rivers Authority has already embarked on plans for some aspects of integrated catchment planning and useful background reviews of some rivers are available as precursors to catchment plans (National Rivers Authority, 1991). The recent initiative of the 'River Restoration Project', which concerns mainly England and Wales, is also welcome.

In developing national strategies for water use and catchment management the question of sustainability must remain paramount. There is little point in promoting wild fisheries in certain rivers if the fishing effort allowed is so great that stocks decline and disappear; or agreeing to water transfer in the upper reaches of a river if this means that the reaches below virtually dry up in summer. The present condition of the upper River Garry in Scotland is a glaring example of just such a decision (Figure 2.3).

Looking to the future, the concepts of sustainable integrated resource and catchment management which have been discussed above are certain to become increasingly important. However, there are likely to be difficulties in moving towards this position, not least of which is the lack of any strong indication on the part of most states that they intend to move in this direction.

8.8 EXAMPLES OF MANAGEMENT

There are many examples of the conservation management of running waters, which, in general, are more difficult to control than standing waters. In general, because of the extensive past usage of streams and rivers and their consequent degradation, both physicochemical and biological, most management schemes are concerned with restoration rather than maintaining the *status quo*.

8.8.1 Streams

Harper's Brook is typical of many lowland streams in England: it has been channelized extensively and some parts have been diverted to allow gravel extraction. The lower parts of this rich, channelized system support prolific weed growths in summer, which enhance silt deposition and obstruct flows – thus necessitating regular dredging and disturbance of these stretches. In response to the environmental and operational concerns over this stream several basic management objectives were established (Smith *et al.*, 1995) to:

- reduce the frequency of weed cutting operations;
- increase the heterogeneity of the substrate, flow and depth;
- provide a more varied channel margin;
- provide a more diversely vegetated riparian zone.

These objectives were achieved by two main management operations. Firstly, 27 artificial riffles were created by removing existing bed material and by introducing coarse gravels and stones over an area about 0.5 m in height and 7–8 m in length, shaped to trail downstream and give a broken flow over the riffle. Secondly, the existing bank profile was modified in several areas to provide a variety of bank slopes, by cutting embayments into the banks, together with a series of shallow berms at or just above water level. Material won in shallowing the bank slope was used to create islands in lakes in an adjacent nature reserve. Shrubs and some trees were planted at various places along the stream bank.

The results of this management have been generally successful. The diversity of invertebrates in the stream has increased and, of course, the stream itself has a much more natural appearance.

The rehabilitation of 24 Danish streams is described by Hansen (1996). Rivers that had previously been deepened and channelled into straight courses were returned to a meandering condition using old maps, where possible, as a guide to the original course, and gravel was introduced to produce riffles and to diversify the substrate and current velocity. Both in large and small natural rivers the distance between riffles in meandering watercourses is approximately 5–7 times the mean width of the stream (Madsen, 1995). The increased gravel provided spawning grounds for salmonids, which increased in numbers, including the rare houting, *Coregonus oxyrinchus*, and the diversity of aquatic macrophytes and invertebrates increased in the pool/riffle situation. Riparian vegetation increased and the watercourses could flood the surrounding ground more easily, aiding water purification.

Where straightened watercourses could not be restored, weed cutting was done by hand to produce a sinuous channel within the existing weed beds, following any deeper parts of the channel, which produced variations in depth and current speed, resulting in biological diversity and increased current velocity (Madsen, 1995).

8.8.2 Rivers

Newall (1995) has shown that the microflow environments of aquatic plants in rivers have an important ecological function which must be considered in the management of rivers. She found that there is less flow variation where there are fewer plants; that trailing submerged plants provide a wider range of velocities and therefore of habitats than do emergent or floating-leaved plants; and that flow velocities are reduced with greater numbers of plants. Thus, stands of aquatic plants and the microflow environments they provide are important in the lotic community and must be maintained as an integral part of river systems.

Newall (1995) has shown that stands of aquatic plants act as refugia for many species of stream invertebrates during periods of disturbance (such as droughts or spates). These refugia later provide a source of recolonizing invertebrates, which move out and re-establish themselves as suitable habitat becomes available. Severe problems arise for the benthic communities when such refugia are disrupted. Armitage (1995) has demonstrated that adequate refugia can be created by artificial means, by providing obstructions in the form of groupings of stones or boulders, etc., but that macrophyte beds themselves can furnish almost as wide a range of conditions. He points out that the old style of manually cutting weeds in streams (Sawyer, 1985) had many management advantages. Certain weed beds would be left uncut and others trimmed to direct the flows from side to side and create areas of faster flow. This maintains habitat diversity and thus a return to such methods would have many ecological advantages, including helping to reduce the impacts of low flows.

8.8.3 Large rivers

Large rivers pose enormous management problems to all concerned – both users (whose activities usually degrade) and conservationists (who are attempting to retain or restore biodiversity). The Missouri River in North America is an excellent case study of the management problems involved with very large rivers (Cieslik *et al.*, 1993). The Missouri River Master Water Control Manual was first published in 1960 and presented the basic water control plan and objectives for the integrated operation of the six mainstream reservoirs in conjunction with various downstream projects. The plan was revised in 1973, 1975 and 1979 as the US Army Corps of Engineers responsible for the scheme became more and more aware of the effects of their operation on the environmental resources found along the river.

Following the effects of a drought that started to affect the river in 1987, pressure from public, state and federal agencies with concerns about the river and its resources became so strong that a review of the Master Manual was initiated in 1989. A major element of the review was an impact assessment of the existing plan and a number of alternate water control plans. The objective was to quantify the effects associated with these water control plans, and an impact assessment methodology was formulated and successfully employed that identified the level of effect associated with the various control plans. This new information is being used in the revised plan for managing this large and important river for the benefit of both humans and wildlife.

8.9 CONCLUSIONS

The biodiversity of a river is closely related to the diversity of the physical (and to some extent, the chemical) habitat available. As the physicochemical

Figure 8.5 A natural obstacle: the Loup of Fintry on the River Endrick in Scotland is a complete barrier to migratory fish.

habitat is simplified and degraded by human activities so does the diversity of the biota reduce. However, within each geographic area there are local and regional variations in diversity within completely natural systems. For example, at local level, a major waterfall on a river (Figure 8.5) will prevent migratory fish gaining access to the river above and this may halve the number of fish species occurring there compared to a neighbouring river with no waterfall.

At regional level, there are major differences between rivers at high latitudes, which are still suffering impoverishment created by the last ice age, and those nearer the equator which were not affected by the ice cap. Thus there is a notable decrease in diversity among riverine fish communities as one moves from low to high latitudes. These waters with simple ecosystems are of value for research and comparison with more complex systems and, as a general rule, the temptation to introduce new species, particularly for angling, should be avoided.

There seems little doubt that the increasing use of rivers for recreational uses will continue and thus there will be greater pressure on this finite resource. However, there is clearly a natural tendency for the public, whatever their recreational demand, to avoid rivers with poor water quality (and few fish), rivers where the landscape is dull and unattractive and, to some extent, rivers where there are substantial conflicts among users. Thus there is undue pressure on those wild rivers that still exist, where the public are attracted by clear water and native fish populations and where the scenery is attractive.

The sensible restoration of degraded rivers will help to improve the balance, for every river where water quality and landscape can be improved will attract people from other high-quality 'wild' rivers and this, together with sensible management policies on each river (e.g. avoiding uses that are clearly in conflict, such as water skiing and angling), will help the move towards the sustainable use of all rivers.

9

Species conservation

9.1 CONSERVATION OPTIONS

The general conclusion from recent species conservation reviews is that protection given to most native aquatic species and their habitats in the majority of countries is inadequate in terms both of the establishment of appropriate reserves and of legislation.

For many groups of plants and animals there is still substantial work to be done in the field of species management. In addition to establishing the status of any group of organisms in a geographic area, much effort must go towards identifying the specific conservation needs of the most endangered species and implementing appropriate measures of protection as soon as possible. As well as habitat restoration, one of the most positive areas of management lies in the establishment of new populations, either to replace those that have become extinct or to provide an additional safeguard for existing isolated populations. Any species that is found in only a few waters is believed to be in potential danger and the creation of additional independent stocks is an urgent and worthwhile conservation activity.

There are two main conservation options available in relation to the conservation of individual species. The first is the protection and conservation management of important habitats (covering all aspects of the life cycles of the species involved). The second involves the production of a species action plan, which gives wide-ranging consideration to all aspects of conservation and acts as a management focus for action by all the people and agencies involved.

The main sections required in a typical species action plan are:

1. Statement of priorities
2. Current action
3. Action plan objectives
4. Proposed action, with lead agencies
   (a) Policy and legislation

(b) Site safeguard and management
(c) Species management and protection
(d) Advisory
(e) International
(f) Future research and monitoring
(g) Communications and publicity

5. Action plan review.

9.1.1 Habitat protection for species

The protection of aquatic habitats of major importance must always be regarded as the prime long-term objective of any conservation programme. The destruction or degradation of habitats by humans has generally been the cause of extinction or decline of freshwater species, rather than over-exploitation. Competition or predation from non-endemic species, introduced accidentally or deliberately by humans, has caused much damage. The classic case is the introduction of the Nile perch to Lake Victoria, discussed in Chapter 3.

When reintroduction of a species into an area where it occurred before is contemplated, the causes of its disappearance must be examined to determine the chances of success. If these causes still exist, alternative sites must be found. Above all, suitable habitat must be present.

9.1.2 Habitat restoration

Obviously enormous damage has been done to many freshwater habitats and the situation is often not easy to reverse – especially in the short term where individual species or communities are severely threatened. In many cases, individual species or potentially unique stocks have completely disappeared. Even where habitat restoration is contemplated, stock transfer (discussed below) could be an important interim measure for some species. However, there are a number of important examples of habitat restoration worldwide and it should be emphasized that habitat protection and restoration are the principal long-term means through which successful conservation of any species will be achieved (see Chapter 5).

Frieberg *et al.* (1994) restored a linear canalized stretch of a Danish stream to a condition resembling its original meandering form. This was followed by an increase in density and diversity of the macrobenthos.

9.2 FISH CONSERVATION

The varied life cycles and individual characteristics of the large number of fish species which occur in fresh waters must be carefully considered in relation to several conservation measures which are available. Characteristics that are especially relevant are:

- **Discreteness.** They are confined within their systems; this leads to independent populations with individual stock characteristics developed since their isolation.
- **Numbers.** Because each population is often confined to a single (often small) aquatic system, within which there is usually significant water movement, the entire population is vulnerable – to pollution, disease, etc. Thus for any species, the number of populations is of far greater importance than the number of individuals.
- **Migrations.** These are a feature of the life cycles of many species of fish and during migration they may be particularly vulnerable. In particular, in diadromous riverine species, the whole population has to pass through the lower reaches of their river at least twice in each life cycle. If the river is polluted, obstructed or has many predators, the entire populations of several species may disappear, leaving the community above permanently impoverished.
- **Life cycles.** Large, slow-growing species and small, very short-lived species are very vulnerable to fishing pressures and can be fished to extinction.
- **Habitats.** Because they are often confined to discrete systems, all the life-cycle requirements for a species must be found within that system. Where this is not the case, species are either migratory or do not establish permanent populations.
- **Ecological niche.** There must be a satisfactory ecological niche within the system to allow population maintenance. This could be disrupted by changes in habitat or the introduction of new species that are predators or competitors.

Habitat conservation is the main way in which fish conservation (and incidentally the conservation of many other species of plants and animals) can be achieved. Already, many wetland habitats are protected both by legislation and active management but this is virtually never for the fish communities there. There is an urgent need for an inventory of fish species already protected in national and international nature reserves across the world (Maitland, 1987a, b). In addition, there is a need to consider those waters that are of major importance to fish species and communities but as yet have insufficient protection. Finally, a number of the most threatened fish species occur in quite restricted freshwater habitats and it is important to protect these waters to avoid the extinction of these rare species.

9.2.1 Translocation

The transfer of stock from one water to another can be done without any threat to the existing stocks, but it is important that certain criteria are taken into account in relation to any translocation proposal.

- The translocation activities must pose no threat to the parent stock.

- The introduction proposals must pose no threat to the ecology or scientific interest of the introduction site.
- The introduction site must be ecologically suitable. In general, sites from which the species concerned has disappeared should be considered to be unsuitable unless the causal factors have ameliorated.
- Ideally, the introduction site should be in the same catchment or the same geographic region as the parent stock; or in the same geographic region as a former stock, now extinct.
- Permission must be obtained from riparian owners or relevant legal authorities, where appropriate.
- Stock may be transferred as eggs, fry, juveniles or adults, but transferring adults may pose a threat to the parent stock.
- Special consideration should be given to the genetic integrity of the stock to be translocated. Once the stock has been defined, maximum genetic diversity should be sought by selecting material widely in space and time. When transferring fish or stripping adults to obtain fertilized eggs at least 30 adults of each sex should be used wherever possible.
- Consideration must be given to avoiding the transfer of undesirable diseases or parasites. Most of these can be avoided by taking eggs only from the parent stock and checking for disease before the eggs or the resulting fry are introduced.
- Notes of each translocation experiment should be kept and details published where relevant.
- The fate of the translocated stock should be monitored.

With many organisms it is possible to obtain substantial numbers of seeds, eggs or young at breeding time without any need to remove adults from the population. Having identified an appropriate water in which to create a new population, the latter can be initiated by releasing transferred stock immediately, or introducing them later at various stages of development (Maitland and Lyle, 1990).

In view of the urgency relating to a number of endangered populations of freshwater organisms, one of the most immediate tasks needing to be carried out is the development of techniques for handling them and establishing new 'safeguard' populations in suitable 'refuge' sites. One of the most difficult aspects of programmes to date has been to locate sites that are suitable ecologically and geographically, that are free from potential competitors or predators and where the owner is sympathetic to the proposals.

An example of where translocation has proved useful is the Arctic charr, *Salvelinus alpinus*, which occurs in only a few lakes in Wales and about ten in England, but many more in Ireland and Scotland – particularly in the north and west (Figure 9.1).

However, it has disappeared from several of its previous waters in England and Ireland. In southern Scotland, where there were previously at

Figure 9.1 The arctic charr, *Salvelinus alpinus*, is a declining species in need of conservation in many countries.

least five populations, only one remains, in Loch Doon (Maitland *et al.*, 1991). The system here is under threat from increasing acidification and current work aims at safeguarding the stock by creating new populations. Thus, eggs were collected from spawning adults in the autumns of 1986–90 and incubated in hatcheries, and the alevins were introduced to a large reservoir in the Scottish Borders (Figure 9.2).

Adults and some reared young were also introduced to a neighbouring reservoir during the same period (Maitland and Lyle, 1990). There is already evidence that fry in the former water have survived and grown well, while young from a successful spawning in the second reservoir have been recovered in the outflow.

A similar procedure is under way with the powan, *Coregonus lavaretus*. There are only seven populations of this species in the whole of Great Britain – two in Scotland, four in England and one in Wales. The largest population is probably in Loch Lomond, where the species is threatened by introduced fish species and other pressures, and in recent years adult fish have been netted here from the spawning grounds in January and stripped to obtain many thousands of fertilized eggs. These have been hatched

Figure 9.2 Megget Reservoir in Scotland has been chosen as an appropriate site to establish a 'safeguard' population of arctic charr.

indoors and the young released into two lochs in the Loch Lomond catchment, previously selected as suitable for new populations, following the translocation criteria outlined above. Sampling in 1991 showed that the fry introduced to one of these sites had grown extremely well and could be expected to spawn within a year or so.

In the longer term it is hoped that all the other rare species will be involved in this project and that even the extinct burbot, *Lota lota*, may be restored to British waters by obtaining stock from waters elsewhere in Europe and reintroducing this attractive fish to some of its former sites. Proposals by one of the authors for this have already been included in the recovery plans collated by Whitten (1990) for endangered species in Britain. International cooperation around the North Sea – by protecting adults of sturgeon, *Acipenser sturio*, and houting, *Coregonus oxyrhynchus*, in the sea and in the few rivers in which they still breed – may also favour the restoration of these species to their previous numbers.

9.2.2 Captive breeding

Captive breeding (in zoos and other institutions) is widely used throughout the world for a variety of endangered organisms, including aquatic ones (Maitland and Evans, 1986). However, for most species it can really only be regarded as a short-term emergency measure, for a variety of genetic and other difficulties are likely to arise if small numbers of animals are kept in captivity over several generations or more. Captive breeding does not seem appropriate in the long term for any of the freshwater fish species at present under threat in the British Isles (Maitland and Lyle, 1992), but is relevant for many of the small species that are severely threatened in other parts of the world. For example, attempts are being made to save at least some of the threatened haplochromine fishes remaining in Lake Victoria by captive breeding in various zoos around the world (Reid, 1990) and this is proposed for other areas with important endemic populations, such as Lake Malawi and Lake Tanganyika (Andrews, 1992).

Short-term captive breeding involving only one generation does have some advantages for a number of species and has already been carried out in several countries. It is especially relevant where translocations are desirable but it is difficult to obtain reasonable numbers of eggs or young because of ecological or logistic constraints. In such cases there are considerable advantages to be gained in rearing small numbers of stock in captivity and then breeding them to obtain much larger numbers of young for release in the wild (Figure 9.3). Because of genetic problems related to inbreeding and loss of genetic diversity it should not be carried out for more than one generation from the wild stock.

In the United States an active programme of captive breeding of threatened fish species is being carried out in premises designed for this purpose at the Dexter National Fish Hatchery in New Mexico, established in 1974 (Johnson and Rinne, 1982). Its main objective is to maintain a protected gene pool of rare fish species, to develop techniques for rearing and maintaining species and to hatch sufficient numbers to re-establish species in their historic habitats, in addition to studying their ecological requirements. Since 1974, more than 20 endangered species of fish have been handled successfully at this hatchery, which is an outstanding example of what is needed in many other parts of the world.

9.2.3 Cryopreservation

Modern techniques for rapid freezing of gametes to very low temperatures have proved successful for a variety of animals, including fish, in order to preserve the material for later use (Stoss and Refstie, 1983). After freezing for many years, and then thawing, the material is still viable. However, the technique is successful only for sperm and, though much research is at present being carried out on eggs, no successful method of cryopreservation of whole eggs has yet been developed. The technique is still at the research stage but recent advances in handling female gametes have indicated that it may be possible to obtain the necessary female information from female sperm (gynosperm) through a selective process called androgenesis.

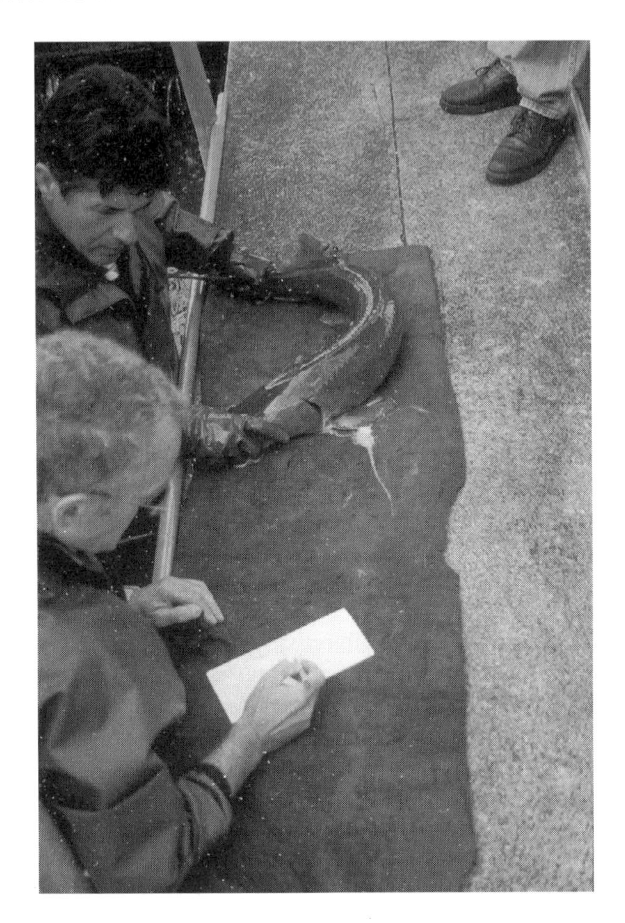

Figure 9.3 Sturgeon conservation in France.

Where a particular stock seems in imminent danger of dying out it would be worthwhile giving consideration to saving at least some of its genetic material through the cryopreservation of sperm. When it is possible to preserve female gametes in a similar way, the technique will have obvious possibilities in relation to the conservation of a wide variety of species. However, even at present, the ability to store male gametes means that some of the genetic problems arising in captive breeding projects may be minimized by the possibility of introducing the original wild genetic diversity into the captive stock at regular intervals.

9.2.4 Monitoring

It is believed that all populations which are sufficiently worthy of conservation must be monitored adequately – at the very least to prove from time to time that they still exist. Ideally, the system used, though it may not be quantitative, should at least be standardized so that comparisons can be made in time, and also perhaps in space in some cases. However, monitoring programmes of this type can be expensive, in terms of both time and finance, and any monitoring programme must be designed to keep both of these to a minimum.

Some basic principles believed to be important are:

- strong legislation to control the importation of potentially disruptive fish;
- controls on the movement and release of fish within each country, on a catchment basis;
- improved fishery management and a complete ban on live baiting;
- active conservation management for threatened species;
- international support for the conservation of threatened species with transnational ranges and migration routes;
- management plans for major sites of all threatened species;
- habitat restoration and management as the major long-term objective, though it may be slow and expensive in the short term;
- translocation of threatened stocks to create new safeguard populations as an important measure, perhaps vital in the short term and valuable in the long term;
- regular monitoring of existing and new sites for threatened species;
- research on poorly known species;
- substantial resources are usually required to implement conservation proposals, which may involve several different organizations;
- once safeguarded, there is no reason to discourage wise, renewable use of any stocks (e.g. for angling or commercial exploitation);
- as well as legislation and active conservation management, an active programme to educate the public (especially anglers) about the value of threatened native fishes;
- regular reviews of the distribution and status of freshwater fish – registers of important stocks should be developed and maintained;
- improved education and interpretation for fishermen, fishery managers, landowners, etc.

In some cases it is possible to take advantage of local circumstances that make monitoring relatively easy. In others, special methods and monitoring programmes must be set up: for this, each site needs to be reviewed and considered on its merits. There is a particular need to investigate the various possible methods of monitoring, especially ones that are non-damaging, cheap and routine to carry out and thus could become part of the duties of local wardening or volunteer conservation staff.

In addition to monitoring stocks of major significance it is essential to carry out regular reviews of the distribution and status of all species in each group of conservation importance (Maitland, 1972, 1992a) and maintain registers of all valuable stocks (Koljonen and Kallio-Nyberg, 1991; Kallio-Nyberg and Koljonen, 1991).

9.2.5 Assessing status

Several difficulties arise in assessing the degree of threat to a species and there are substantial anomalies in the conclusions that can be reached using different methods and also in comparing various groups of plants and animals. The most reliable methods are objective ones, which stress the importance of the number of individual stocks of a species and whether or not these stocks are increasing or decreasing.

As an example of the kind of anomaly that can arise it is useful to compare the conservation status of fish species that have been identified as threatened in Great Britain (Maitland and Lyle, 1991) but are not included in Schedule 5 of the Wildlife and Countryside Act (this includes plant and animal species that are believed to be in need of protection in Great Britain). For example, the least threatened of the ten fish species identified as threatened by Maitland and Lyle (1991) is the Arctic charr, which, though in decline, still has about 200 populations in the British Isles. In comparison, the crested newt, *Triturus cristatus*, which is listed under Schedule 5, is also supposed to be in decline but is estimated still to occur at some 18 000 sites in Great Britain (Whitten, 1990). The smelt, *Osmerus eperlanus*, which has declined substantially in recent years, is still fished commercially at several of the 20 or so remaining sites. The pipistrelle bat, *Pipistrellus pipistrellus*, however, included in Schedule 5, is regarded as vulnerable in Great Britain and in Europe, yet in Britain alone there are thousands of breeding colonies of this common species.

Fortunately, there is provision within the existing legislation for a 5-yearly review of schedules by the successor bodies to the Nature Conservancy Council. This process allows new species to be added or species not at risk to be deleted. However, one problem related to conserving fish compared with other organisms is that many more species are exploited commercially or by anglers. Often, completely different methods of capture and exploitation are used in different countries and species regarded as important in one country are ignored in another. Attitudes and legislation in relation to fish conservation and stock management vary widely across Europe.

The general picture around the world is that the conservation of most aquatic organisms has been sadly neglected, especially compared to the attention that has been given to terrestrial birds, mammals, plants and some invertebrates. There are a few exceptions to this, the outstanding one being the United States. Here, an Endangered Species Act (which includes many aquatic species) became law in 1973 and has been updated several times since. The Act is implemented within the United States by the US Fish and Wildlife Service and the three main areas of emphasis are in listing the species under threat, protecting the habitats concerned and actively managing to restore populations that have undergone serious decline (Williams and Miller, 1990).

However, in the last few years more and more attention has been given to fish conservation internationally and several symposia have been devoted to this topic (e.g. Maitland, 1987c; Le Cren, 1990). Several reviews of the conservation status and needs of freshwater fish in large geographic areas have been prepared (e.g. Maitland, 1986, 1992b; Pollard *et al.*, 1990; Skelton, 1990; Williams and Miller, 1990; Crivelli and Maitland, 1995) and it is hoped that these initiatives will lead to practical conservation management schemes to save threatened fish species and restore important habitats on a worldwide basis.

9.3 BIRD CONSERVATION

Many techniques have been developed for managing aquatic birds, particularly in North America. These have mostly been associated with hunting activities, e.g. Ducks Unlimited, and occasionally with commercial activities like farming eider ducks for their down. The techniques used are applicable to the management of birds for conservation purposes. They consist principally of either improving reproductive success, providing secure roosts or improving feeding conditions. The former has been done generally by providing artificial breeding sites but can only be justified when there are not enough natural sites available or when nesting success is low for some other reason. Care must be taken not to draw concentrations of birds into insecure situations where they may be vulnerable to predation. The structures used should be compatible with the concept of nature conservation and aesthetically pleasing. The use of materials such as oil drums, washing tubs and plastic boxes to construct nesting places should be avoided! Improving food supplies is generally linked to wetland habitat management (Chapter 6) or farming for waterfowl. Efforts are being made to persuade hunters to replace lead shot in their shotgun cartridges with steel or plastic shot, to reduce saturnism in water birds, and similarly for anglers to find substitutes for lead weights (see section 3.1.4). The following account outlines some of the techniques used to date.

9.3.1 Tree-nesting birds

Herons, cormorants and various ducks, which nest in holes in trees, are severely affected by tree felling and the removal of dead wood from the forest. Cormorant colonies often kill the trees in which they are nesting with their own droppings.

Meier (1981) constructed artificial nesting sites beside a colony of *Phalacrocorax auritus* in decaying trees. Triangular platforms, with sides of about 55 cm, made of laths, were supported on poles at a height of 3–8 m above the water. The sides of the platforms protruded to provide perches and Meier considered that filling the platforms with nesting material helped to gain acceptance by the birds. Sandilands (1980) and Meier (1981) found that great blue heron nested in similar structures in flooded forests on poles 9–11 m long. On the other hand Wiese (1976) established colonies in Louisiana of up to 430 pairs of great white egrets, on large platforms up to 7 × 35 m in area held 2 m above the water. The framework of the platform was covered with bamboo poles and pieces of bamboo were distributed on this to provide nesting material.

Hafner (1982) describes a long-term technique to establish a mixed nesting colony of herons (*Egretta garzetta, Bubulcus ibis* and *Nycticorax nycticorax*) in the Camargue, France. Many of the existing colonies are threatened and it was decided to try to establish a colony in a protected area. An island 40 × 80 m was constructed by digging out a moat, which filled with water from a nearby canal, in order to give security and protection from predators. In 1971 about 5000 shoots of mainly *Fraxinus angustifolia*, plus *Populus alba, Ulmus minor* and *Salix* spp., which are the common constituents of the nesting woods in the Camargue, were planted on the island. By 1977 it was considered that the wood had reached the structure desired and in February 1978 an aviary was built containing two dead trees in which artificial nests, mimicking real ones, were built. Ten herons were introduced into the aviary and 50 white polystyrene silhouettes of egrets were placed in the trees on the island. As the presence of large quantities of dead twigs is important to initiate nesting colonies (Valverde, 1955; Hafner, 1977) several piles of dead branches were placed on the island. In May 1978 a pair of egrets bred in the aviary and in November 1978 little egrets and cattle egrets started to roost on the roof of the aviary. The vicinity was used variably as a roost over the next 2 years and in 1980 three pairs of wild little egrets bred on the roof of the aviary. Five of these had been taken from wild nests in 1978, when 10 days old, reared in the aviary and released in spring 1979. In 1981 there was a spectacular invasion of breeding birds into the newly created wood and about 900 young were raised in 225 nests of little egrets, 35 nests of cattle egrets and 56 nests of night herons. This study is an excellent example of the design of management techniques based on knowledge of the behaviour of the species and their habitat requirements for breeding. It is environmentally sensitive and a model for all.

In Scandinavia, old cart wheels are sometimes placed horizontally on the top of long poles to provide nest supports for ospreys, *Pandion haliaetus*. These in some ways resemble the structures placed on roof tops for nesting storks. They serve as replacements for old nesting trees that can no longer support nests, and can be sited near them for this purpose. Postupalsky

(1978) found that losses due to nests being blown down were only 6% of total mortality in artificial nest sites and accounted for 57% of mortality in natural sites, showing that this can be an important factor in mortality in this species.

Several species of ducks, such as whistling ducks, *Dendrocygna* spp., wood ducks, *Aix* spp., and goldeneyes, *Bucephala* spp., nest in holes in trees and numerous designs of nesting boxes have been described to accommodate them (Lumsden, 1982). No standard box has been found to be particularly attractive to any one species and different populations of the same species may have different preferences. The principal points of design are as follows.

The main criteria are to have an entrance hole of the right size – 9–20 cm in diameter, depending on species – and a suitable cavity to lay the eggs. This should have a layer of wood chips in which the duck can make a depression to accommodate the eggs as she does not collect nesting material and it is only her own down that is added to the material already in the nest box. The floor should be provided with drainage holes as a precaution against rain or snow driving into the box. Boxes are best fixed to trees with wooden dowels so as to avoid nails or screws, which can finish in the interior of the wood as the tree grows, and they should be secured so that there is no movement. The height above the ground is not too important but it is better to be more than 3 m to foil curious humans. The box should be sited near to a branch, which can act as a landing perch for the ducks. Several authors report the best occupation success in boxes near to water and with the entrance directed towards the water. The side of the interior of the box, below the entrance hole, should be rough enough for the ducklings to climb out or a ramp must be constructed. Nesting boxes often attract predators and they must be sited with this in view. Avian predators, such as crows and jays, can be foiled by constructing a porch on the box, and a cone-shaped metal sheet placed around the tree or supporting post and protruding 45 cm will prevent most mammalian predators climbing up. More agile mammals, such as squirrels and monkeys, are more difficult to thwart.

The power of attraction of nesting boxes is shown by the goldeneye, *Bucephala clangula*, which was only known as a wintering bird in Scotland. In an attempt to obtain a nesting population, boxes were placed at a site in the Scottish highlands where the latest birds lingered in the spring before migrating northwards. This attracted birds to nest in the boxes and there is now an established nesting population. Where old trees with suitable nesting holes have been lost from an area, nesting boxes are a well proven method of maintaining populations of tree-nesting ducks.

9.3.2 Ground-nesting birds

Boxes at ground level, or sunk slightly in the ground, having a tunnel entrance, will be used by shelduck, *Tadorna tadorna*, in place of their habitual burrows. Makeshift nesting places can be made from a few bales of hay stacked so as to leave a nesting space underneath and a tunnel entrance (Jenkins, personal communication). In Iceland and Baffin Island, where the down is collected commercially from nests, eider ducks, *Somateria mollissima*, readily accept ground nesting boxes, which give shelter and greatly reduce predation on eggs by gulls (Clark *et al.*, 1974).

Many dabbling duck species will utilize rectangular or tent-shaped boxes placed in vegetation on the ground. Oleinikov (1971) described various structures, made from materials present on the site, which are more appropriate for use in nature reserves. An area about 45 cm in diameter was cleared in dense shrubs and a hollowed-out turf was placed within it to accommodate the nest. The surrounding stems were then bent over it and tied together to form a protective tent. He made similar structures of reed in reed beds and also constructed hut-shaped shelters of the same material. Where there was danger of a rise in the water level he placed these on floating bundles of reeds, which remained buoyant until after the breeding season. Structures built of natural local material in this way have the advantage that they do not leave untidy litter on the site when they deteriorate.

A flask-shaped basket woven out of wicker has been used for centuries in Holland as a nesting site for ducks. These are placed on their side and are usually fixed to poles a metre or so above water to give security from terrestrial predators. The aperture protrudes on the lower side to provide a landing platform for the ducks. These baskets are readily accepted by mallard and are in wide use in Europe and North America. Various types of nesting basket have been used, both open and covered (Lumsden, 1982), made of a variety of materials ranging from wire netting and wood to plastic containers, oil drums and galvanized wash tubs! Even Canada geese, *Branta canadensis*, nested in these last filled with sawdust and a layer of straw (Brakhage, 1965). It is clear that such materials should not be used in environmentally sensitive areas if they are not to look like a rubbish dump and, with a little thought, aesthetically pleasing structures incorporating the same principles can be made from natural materials such as wood or reeds.

9.3.3 Nesting rafts

Where there are no islands and it is not possible to build them, nesting rafts to a limited extent provide a substitute habitat for ground-nesting waterfowl. These rafts have the advantage over Dutch nesting baskets that they are not influenced by fluctuating water levels, so that ducklings and adults always have easy access. If designed and planted with care they are also more pleasing to the eye. They have been used extensively in Britain, particularly in conjunction with the rehabilitation of artificial water bodies such as gravel pits.

Figure 9.4 Rafts for birds: the shingle raft in the foreground is for terns, with a duck nesting raft beyond, which at the time of the photograph held four nests of mallard and two of tufted duck.

Among the first rafts to be used were those designed by Harrison (1972). These consisted of two metal floats 250 × 50 × 50 cm attached to an angle-iron frame strengthened by wooden cross-beams. A metal grid was fixed over the cross beams and clumps of rush, *Juncus* sp., were planted on this. Wooden boxes were built around the floats and were filled with 15 cm of soil to accommodate further vegetation. Since then, rafts have been built of a variety of materials, depending on what was available (Figure 9.4).

Polystyrene, wrapped in plastic sacks to protect it, has been found to be an excellent material for floats that has the advantage of being cheap and very light to handle. If metal drums are used as floats it is better to have several small ones: these are more easily replaced and the raft will not sink if one is pierced. Plastic-covered wire netting and chain-link fencing are now freely available and are excellent materials for covering the rafts, being long-lasting. The rafts should be ballasted to float only a few centimetres above the water level and ramps with a rough surface should be built along either side of the raft by which ducklings can climb on to it. Metal loops or rings fastened on to the corners facilitate towing into the position where the rafts are to be moored. They must be anchored from opposite sides to prevent them turning with the wind, using wire ropes attached to weights such as blocks of concrete.

For use by ducks, the rafts are covered with wetland vegetation (e.g. *Juncus* spp., *Carex* spp, *Cyperus* spp.) to provide cover. Ground nesting boxes can be added if desired. To give a natural look, earth can be added to produce a higher elevation in the middle of the raft and construction to give an irregular or oval outline adds to the natural effect. A manageable size is about 5 × 2 m. When not in use for nesting, rafts also provide safe roosting sites.

Rafts have also been used successfully for nesting terns and waders, in which case they are filled with pebbles with no vegetation. A vertical board should be fitted along the sides to prevent the young from falling in the water. Pieces of guttering 30 cm long can be inverted on the gravel to provide shelter for the chicks from avian predators.

9.3.4 Farming

Wintering geese, swans and some ducks, such as wigeon, feed extensively on flood meadows and roost in the security of nearby water bodies. Disturbance by man in either habitat can reduce or stop their use and winter grazing by farm animals may limit the food supply. In Western Europe grazing by cattle, sheep and horses is used to provide the necessary short Graminae for grazing birds (Thomas, 1982) and short grassland habitat for nesting limicoles (Bientema, 1982). As such grazing is selective, certain unpalatable tussock-forming plants, such as *Deschampsia caespitosa* and *Carex riparia*, form nesting cover for ducks and waders. When the density of cattle is higher than four per hectare, the nests of lapwing, *Vanellus vanellus*, are destroyed by trampling. Bientema (1982) considers that adult cattle should be used as grazers during the bird-nesting season as they are less active than horses and therefore trample fewer nests. However, horses eat more than twice as much vegetation per individual as cattle and a lower density of horses therefore achieves the same grazing impact. Owen (1982) experimented with the management of 128 ha of saltings used by white-fronted geese, *Anser albifrons*, in south-west England. He stopped cattle grazing on 30 September, which prevented winter disturbance, and reduced the autumn grazing pressure on the sward to leave more vegetation for the geese. As a result the carrying capacity for geese was almost doubled.

In the United States it has become common practice to plant arable crops – maize, wheat, rice etc. – on waterfowl reserves to provide winter food. Some of the crop is left standing and that which is harvested is spread on the ground as food becomes short. This technique was used very successfully at Horseshoe Lake to attract and protect wintering Canada geese, which were being heavily shot out on the neighbouring Mississippi River. However, this technique must be viewed with caution, as the majority of wintering mallard and Canada geese in the USA are now dependent on these farmed refuges, which means that any change in policy could be disastrous for them.

9.4 INVERTEBRATE CONSERVATION

Relatively little research has been done on the conservation of aquatic invertebrates and few management activities have been directed at them. In general terms, if the habitat is in good condition the related species will survive or where a site is rehabilitated (see section 9.1.2) those insect species with flying adults will rapidly colonize. The arrival of those forms without a flying stage depends on the particular circumstances: the presence or absence of inflow streams, transfer with introductions of fish or plants by humans and transfer on the feet or feathers of aquatic birds are all possibilities. Rees (1965) records the dispersal of molluscs on the feet of wading birds. Proper targeted management depends upon knowing the precise ecological requirements of a species, particularly factors that may be limiting its presence. With rare species it is important to know, by adequate survey and monitoring, what makes them rare, where they occur and in what numbers. Such knowledge is often lacking. A few examples of the conservation of individual invertebrate species and groups are given below.

The Odonata have received more attention than most groups because they are attractive. The nymphs of many species are found in standing water pools and provision of such habitats will be sufficient to ensure the presence of a number of species. Pretscher (1983) studied the requirements of 24 species and was able to allocate them to four different pond types differentiated by the stage of vegetational development. At altitudes below 1200 m, in Natal, Samways (1989) found that 78% of the species present were only found in farm dams. He attributed this to the stability of this habitat compared with the hostile natural habitats, which dry out in winter and are subject to scouring torrents in summer. Moore (1964) found that in woodland ponds, it is important to cut the trees on the south side of the pond to provide open sunny areas where the adult dragonflies can fly.

Studies by Steytler and Samways (1995), carried out in a pond created by damming a stream, showed an increase from 12 to 26 species of dragonfly in relation to the new habitat with the loss of only one species. A wide variety of biotopes became available, from floating aquatic vegetation through emergent short grass to tall trees. Correlations were determined between the more abundant dragonfly species and environmental variables: eight taxa had an affinity for sunshine and two for shade; five species were associated with running water and four with still water; seven species correlated with short grass, others with long grass and two with forest; one species correlated with one species of grass, *Setaria megaphylla*. Studies of this type are essential if we are to advance to more precise techniques of managing individual invertebrate species. Moore (1991) points out the merit of having different successional stages of vegetation to maintain maximum dragonfly diversity.

One mayfly (Ephemeroptera) is listed in the IUCN Invertebrate Red Data Book – *Tasmanophlebia lacus-caerulei*, which only occurs in five small lakes near the summit of Mount Kosciusko, New South Wales, Australia (Tillyard, 1933). Fortunately all five lakes occur within the national park there; water quality is being studied and a monitoring programme for this rare species is now in operation.

The large copper butterfly, *Lycaena dispar*, became extinct in England and was reintroduced into Woodwalton Fen, where it is maintained by management of the greater water dock, *Rumex hydrolapathum*, which is a principal food plant of the caterpillar. Duffey and Morris (1965) and Duffey (1968) showed that the critical factor is to have sufficient isolated young plants, and management is directed to this end.

The freshwater crayfish, *Austropotamobius pallipes* and *Astacus astacus*, have been eliminated from many traditional sites by a fungal disease, *Aphanomyces astaci*, which spread throughout Europe. Work has been carried out in Finland on rearing native species in hatcheries and reintroducing them in the wild. It has been found that, to be successful, the water to which they are introduced must be completely free of crayfish, as the disease cannot maintain itself in their absence. Otherwise, the introduced animals will be infected from the residual crayfish population (Lindqvist, personal communication). Measures are in force in several countries in Europe to control fishing for crayfish, with size limits, regulations regarding fishing methods and close seasons. It is important to prevent pollution of the sites where they still occur and the introduction of non-native crayfish must be avoided. As several alien crayfish, mostly from North America, are being used in aquaculture in Europe, it is difficult to prevent escapes, which may subsequently establish populations and eliminate the native species.

Spengler's pearl mussel, *Margaritifera auricularia*, was once widespread throughout Europe, where it occurred in wide, deep rivers with clean water. It has declined over the centuries due to collection by humans and, more recently, has been lost from most rivers owing to pollution. Altaba (1990) studied one of the last, declining populations in the River Ebro in Spain. It is highly endangered from water extraction, engineering works and pollution and a list of recommendations for its protection have been accepted by the administration of the Ebro Delta Natural Park. These are:

- restricted collection by permit only;
- research on the *Margaritifera* population;
- protection of the fish hosts of the glochidia;
- protection of key habitats;
- education of the public;
- establishment of captive breeding colonies, which can ensure high survival of glochidia (Unionid larvae have metamorphosed *in vitro* – Isom and Hudson, 1982);

Figure 9.5 A common frog, *Rana temporaria*, in its natural wetland habitat.

- reintroduction;
- water quality control.

If this list is put into effect it should be possible to save this population of *Margaritifera*, but it depends on close cooperation between conservationists, the authorities and land owners and the general will to succeed.

9.5 AMPHIBIAN CONSERVATION

Wetlands are valuable breeding sites for many species of amphibians, the main requirement being that parts of the wetlands retain a minimum of 10–20 cm of water during the spawning period and larval stages (Figure 9.5).

To investigate conservation measures Laan and Verboom (1990) compared the use of 38 old and 39 new pools in the Netherlands by eight species of amphibian: four anurans and four Caudata. The number of species using a pool was linked positively to the age of the pool and the presence of a species was related to the distance from the nearest pool where it already occurred. The proximity of a wood proved to be an important factor in relation to the presence of amphibians in both new and old pools. Woodland was considered to be an important element in the continuity of the landscape and as habitat for Amphibia. These are important points to be considered when providing new habitat. Access is important, particularly if migrating amphibians must traverse roads to reach their spawning grounds. This can lead to high mortalities and in sensitive areas walls and underpasses are constructed to direct migrating animals safely under motorways.

Schlupp and Podloncky (1994) successfully translocated toads, *Bufo bufo*, which have a strong fidelity to their breeding sites, to an artificial pond to prevent road mortality. A barrier was made to prevent the toads from crossing the road, the toads were collected in pitfall traps and transferred to the substitute pond. The majority of the adult population became attached to the new pond after 2–3 years. The migration towards the old site dropped by 15% in the fourth year and to less than 1% thereafter.

9.6 MAMMAL CONSERVATION

Mammals associated with fresh waters have suffered from loss of habitat, from excess hunting pressure for food or pelts and from pollution. The beaver, *Castor fiber*, has declined in parts of Europe and Russia under these pressures and has been successfully reintroduced into protected areas following captive breeding programmes (Grzimek, 1965; Djoshkin and Safonov, 1972). The essential habitat feature is that the area of introduction contains the necessary trees, such as *Populus* spp., *Salix* spp. and *Alnus* spp., which are used as food plants. The otter, *Lutra lutra*, has also declined greatly in Europe but in this case the principal cause has been the bioaccumulation of toxins, particularly dieldrin, PCBs and mercury, (Skarén, 1988; Mason and Madsen, 1990). While traces of these remain in a river, reintroductions will be ineffective. Where river waters are clean, it is essential to obtain the cooperation of local anglers, who are often opposed to the introduction of a fish predator.

On the other hand several species such as muskrat, coypu and American mink have established themselves at several places in Europe, having escaped from farms where they were reared for the fur trade. They have caused problems such as the destruction of aquatic vegetation, the undermining of dykes, predation on fish and aquatic birds and competition with endemic aquatic mammals. The desired management here is to control or eliminate the invading species by trapping, or other means, but it is difficult to achieve success when a species is well established.

10

Future prospects

10.1 INTRODUCTION

Analysed against the many problems facing the conservation of natural and semi-natural habitats on a world scale, future prospects are laden with difficulties which it would be unrealistic to ignore. The continued demographic explosion in most underdeveloped countries is leading to greater and greater pressures, which are eliminating valuable natural areas on an accelerating scale. Fresh waters are particularly at risk from demands for clean drinking water, irrigation and hydroelectric power and from being drained for agriculture or control of vectors of disease. Even in developed countries, important sites are still being lost to engineering projects on rivers and lakes and drainage of wetlands continues where pressure groups do not exist or are unable to resist strong political pressures that disregard environmental issues.

For example, to cite but one country, in France the construction of three reservoirs is going ahead on the River Loire (Figure 10.1), the only major French river which is still in a near-natural state, and Fournier and Wattier (1979) reckoned that 100 km^2 of wetland was being damaged or destroyed there each year.

This is in spite of the fact that French agriculture produces ten times the amount of food required by its own population and Europe as a whole has an embarrassing agricultural overproduction.

Thus the major problem for the future is a political one, which is outside the scope of this book but which nevertheless must be faced if conservation is to succeed. We now have adequate experience and research-based information with which to advise planners and politicians and the important thing for the future is therefore a good two-way communication between ecologists and planners and politicians. This will only be productive if the latter are receptive to ecological arguments and for this, one of the principal requirements is environmental education of the public, which will then

Figure 10.1 The River Loire, one of the few largely natural rivers remaining in France.

diffuse through the whole of society. The media, particularly television, can play an important role in developing public awareness.

10.2 PEOPLE AND EDUCATION

It is important that local people and traditions are integrated into conservation projects, so that not only are the habitats, flora and fauna protected, but local people continue to benefit from the use of the area and the sustainable exploitation of its resources. This gives a sense of the value of the area to such people. Traditional activities such as fishing, reedcutting and hunting have often been going on for centuries beside the conservation interest and it is important not to create conflict situations but to identify the benefits and determine means of compatibility. This concept of multiple sustained use has been successfully developed by the Worldwide Fund for Nature in such projects as that at the Kafue Flats in Zambia. The same principle has been applied in UNESCO's biosphere reserves.

A most urgent task for the future is to put the conservation of fresh waters into effect against the background of enormous pressures generated by the human population explosion. To convince planners and politicians, more research is required to give precision on the intrinsic values of these systems by studies on their functioning. Schrader-Frechette and McCoy (1994) consider that ecological science has been largely explanatory and has failed to be predictive, an essential to management. They propose that prediction can best be arrived at through case studies. Such research will also contribute to refining management techniques in protected areas and should continue to be complemented by experimental management studies.

It goes without saying that a considerable expansion in training facilities for research scientists and managers is required, particularly in underdeveloped countries, if conservation aspirations are to be put into effect. Strong governmental and institutional support for international associations such as IUCN, UNESCO and FAO, which play a major role in such activities, is essential.

10.3 SUSTAINABLE RESOURCE USE

In developing national strategies for water use and catchment management the question of sustainability must remain paramount. For example, there is little point in promoting wild fisheries in certain rivers and lakes if the fishing effort allowed is so great that stocks decline and disappear; or permitting water transfer in the upper reaches of a river if this means that the reaches below virtually dry up in summer.

Freshwater fishes are an important source of human protein on a worldwide basis and the annual commercial catch is many millions of tonnes. In addition, there are important subsistence fisheries in several countries (the catches usually unknown), while sport fishing, in some countries at least (e.g. the United States), exceeds the commercial catch and is now a major source of income to the economy of many, otherwise poorly resourced, areas.

Many commercial freshwater fisheries using traditional (but often inefficient) methods have shown themselves to be sustainable – since they have existed successfully for hundreds of years (Maitland, 1994). In fact, their success has been due mainly to the underuse of the stocks concerned. Most modern fisheries have shown themselves to be unsustainable, because the use of electronic location methods, modern fisheries gear and boats and the ability to freeze and store enormous quantities of fish has created the ability to decimate populations within a few years.

Thus, the successful management of freshwater fish populations to allow their sustainable use must rely on a sensible mixture of annual scientific information about each species and the extrapolation of this, using appropriate statistical models, to determine the allowable catch for the following year. This must be supported by appropriate laws and guardianship. There must also be equitable policies where international waters are concerned and a realization that, in the long term, the status of freshwater fish populations is dependent not only on the quality of the water in which they live but also on the land use and other activities by humans in the catchment which it drains.

The same argument applies to commercial fisheries for crustaceans and molluscs as well as hunting for water birds and fur-bearing mammals.

10.4 CATCHMENT MANAGEMENT

Although the importance of whole catchment management has been realized for many years, scant attention has been paid in most countries to its practice. Water can be conserved only by interrupting the hydrological cycle to make more water available, or to make it available for longer periods, or to use less (Buchan, 1963).

10.5 INTEGRATED WATER USE

The regular increase in the quantity of water required for domestic and industrial (including agricultural) purposes shows no sign of lessening, making it necessary to consider the whole question of water resource and supply on a more extended integrated basis. These increasing demands continue, but integration is poor. Water conservation means the preservation, control and development of water resources to ensure that adequate and reliable supplies are available in the most suitable and economic way while safeguarding all legitimate interests, including the natural environment (Figure 10.2).

Historically, only extremely pure waters were considered as potential water supply sources, and this emphasis on quality led to priority exploitation of easily available supplies, with many impoundments in upland areas. The shortage of water and the improving means for treatment of contaminated water (Figure 10.3) have meant that, increasingly, quantity and not quality is now often more important; sources that would not have been considered in the past are now accepted and treated for public supply.

With so many waters in most parts of the world there should be enough to serve most needs, but present national and international planning is inadequate and there are many problems and conflicts to be resolved. One solution is the creation of a national framework and policy for the major lakes, rivers and wetlands of the world. This would look at their distribution, quality and value in relation to demands on them. Some are clearly essential to a major need (for example, hydroelectricity or water supply) and this must be the over-riding factor – though other uses may fit in with this. Other waters may be of such high wildlife importance that their conservation needs are paramount. Several of the largest waters are so important nationally or internationally that each requires an individual management plan in order to reconcile the needs of potential users with the overwhelming importance of maintaining the quality and value of that water (see Lake District National Park, 1993). Loch Lomond, Scotland, is a topical example, where the pressures have become so great that the

Figure 10.2 Integration of industrial needs and wildlife conservation are explained during an educational tour of the Tongland Power Station in Scotland.

Figure 10.3 A reed bed for sewage treatment. Many systems for treating pollution are changing to such 'green' solutions, which have many benefits for wildlife.

government appointed a working party to produce a management strategy (Loch Lomond and the Trossachs Working Party, 1993). In all such cases, users must be prepared to accept compromises and even give up claims somewhere – if not on one water, then on another.

Clearly therefore, national strategies must be developed to integrate the present and future needs of society in some form of constructive management system. Such strategies can only be developed in relation to appropriate interactive computer databases, which can hold all the information necessary to making objective management decisions, regarding the wisest use or uses for each fresh water at both local and national levels. Substantial amounts of information need to be collected and collated in such a way that they are easily available for analysis. Moreover, the various uses must be objectively assessed so that their impacts both on the water body itself and on each other are understood. While it is obvious that, say, noisy power boating and contemplative angling are inimical to each other, this is not necessarily true of, say, water supply and bird watching.

The value of such a database to local and other authorities in planning regional strategies is obvious and this value relates not only to the uses of water bodies in any area but also what activities may be permitted within a catchment. However, almost by definition the scheme can only realize its full potential when extended to national and international level. Only then will it be possible to prioritize and integrate fully the needs of all users and waters.

Future work is required on techniques for showing the value of freshwater habitats in quantitative terms, not only for fisheries or the extraction of peat for fuel, which have a clear economic value, but for such things as gene pools, interesting plants and animals and scenery.

10.6 CLIMATE CHANGE

Present estimates of the rates of global warming are variable but there is general agreement that it is taking place. At present it is not possible to say with certainty which areas will get wetter and which drier, nor which will warm up the most, but there is a consensus that some wet areas will become wetter and some dry areas drier. As more precision is obtained through research it is important to determine which fresh waters will remain viable, although possibly changing in character, which will dry out and where new ones will appear. However, as the most rapid warming will probably be at the polar regions, the most evident early effect will be the invasion of the tundra by the northern forests, with serious consequences for the water birds that nest on the open tundra of the Arctic. With rises in sea level due to the melting of the polar ice, it is certain that many coastal wetlands will become saline or brackish and that coastal drainage will be impeded. If evapotranspiration increases with increasing temperatures, then the amount of runoff into rivers will be reduced, giving diminishing flows and reduced supplies to lakes and wetlands. In the long term, global warming is probably the greatest threat to the integrity of many fresh waters, but in the short term the threats discussed in Chapter 3 are the immediate problems.

It is clear, then, that all those concerned with the environment and attempts to manage parts of it are faced with a dilemma. However, though there are still many uncertainties in attempting to assess the nature of future ecological changes this should not put scientists off discussing and debating the issue (Toha and Jaques, 1989; Regier and Meisner, 1990). Looking at all sides of this multifaceted argument can be constructive in demonstrating gaps in our understanding of the effects of climate on aquatic systems and in particular on current knowledge of the biogeography and physiological ecology of aquatic organisms – two scientific disciplines that are highly relevant to any assessment of the effects of climate change.

In addition, although there are many uncertainties in the situation it can be looked upon as a giant global experiment, of which scientists should take advantage by accumulating baseline data and setting up projects that will help to understand much more about aquatic systems, predict changes and manage them. 'Innovative monitoring can provide data for enhanced predictability of ecosystem change' (Kennedy, 1990). In the case of many fisheries, management policies may need to be modified (Healey, 1990).

Tonn (1990) has emphasized the need to consider the importance of factors such as isolation, extinction and colonization, as well as simply temperature, in making climate change assessments. He suggests a community-level framework, which can organize accumulated knowledge of fish assemblages, identify causal processes behind patterns and focus research needed for the management of fish assemblages in the face of anticipated changes in climate. Thus it may well be that we have to move away from the emphasis on predictive models (DeAngelis and Cushman, 1990) towards more positive forecasting by analogy (Glantz, 1990). There are valuable lessons to be learned from the careful study of retrospective case histories, as Soutar and Isaacs (1974) have shown.

Since the advent of fears about global warming there has been a marked change of attitude to monitoring programmes. Long-term studies that were previously criticized by experimental scientists are now proving of major importance in studying trends. Funding that was being removed from a number of important projects has now been restored and new monitoring programmes have been instigated.

If the forecast changes for global warming are correct there will be serious consequences for many rarer species. It has been suggested that species affected by global warming will respond by genetic adaptation and so escape most of the worst effects. However, it seems to be generally agreed that the rates of climate change are likely to be so fast that species will not be able to respond quickly enough.

10.7 LEGISLATION

The amount of legislative protection given to freshwater areas naturally varies from one country to another. Hassan (1993) emphasized that, in the great majority of countries, the legislation relative to wetlands is insufficient. Often the laws applicable to wetlands are related to the wish to see them disappear by drainage and other means, and can be exploited by those who have no interest in their conservation or rational use. The existing legislation must be revised and strengthened to take into account the protection of important ecological zones. It is important that laws are sufficiently flexible to allow, for instance, discrimination between protection for an area such as a National Park and more flexible legislation for zones with multiple uses. Clearly, too, there must be provision for the control of harmful activities outside a protected area but still within its catchment. This must include mechanisms for coordination between local, regional and national legislation, particularly in countries where the power of decision making is strongly decentralized, which impedes overall conservation planning.

As well as biological research, further research on appropriate methods of legislation is required. It is clear that, generally, standards of control of threats to the integrity of fresh waters, such as pollution, should gradually be strengthened on a worldwide scale. Attempts have been made to develop some policies at an international level, for example to control fish introductions (Ryder and Kerr, 1984). It is important to realize that many water bodies traverse the boundaries of several countries and this demands compatible policies and legislation.

10.8 INFORMATION

Provided that satisfactory legislation is available, it is important to know where the sites of greatest ecological importance are situated. The first attempt to list sites of international importance was Project Aqua, drawn up during the International Biological Programme by Luther and Rzoska (1971); this was intended as a source book of inland waters proposed for conservation and contained more than 600 sites. This was followed by a directory of western palaearctic wetlands complied by Carp (1980) for UNEP and IUCN, and most recently, IUCN has published a directory of wetlands of international importance (Anon., 1990). The latter incorporates information from directories of neotropical wetlands (Scott and Carbonell 1986), African wetlands and shallow water bodies (Burgis and Symoens, 1987) and Asian wetlands (Scott, 1989).

Information available on wetlands is therefore fairly comprehensive, although there is considerable variation in the quality and amount of data available from different sites. The Ramsar Convention (Anon., 1971) has been an important instrument in integrating the protection and management of wetland sites throughout the world; by 1990 55 nations were signatories, each with varying numbers of sites (Dugan, 1990), and by July, 1996, 93 nations had designated 808 sites covering nearly 54 million ha (Anon., 1996). Collated information at the international level for open waters is less comprehensive but is often well documented at national level. It is necessary here to fill the gaps and carry out an overall evaluation. To convince planners and politicians, more research is required to give precision on the intrinsic values of these systems by studies of their functioning. Such research will also contribute to refining management techniques in protected areas and should continue to be complemented by experimental management studies.

Politicians will not move without public pressure and it is therefore essential that the public should be informed about the values of freshwater systems and the threats that face them. Even in developed countries the person in the street is often ill-informed and therefore ignorant of the needs to conserve these ecosystems. There is an urgent need to develop educational programmes on a world scale to remedy this situation.

10.9 CONCLUSIONS

In spite of the many disasters and extinctions that fresh waters and their biota have faced around the world there are reasons for optimism. In many countries both the government and the non-government organizations are taking initiatives that bode well for the future – at least in some areas of aquatic conservation.

For example, in Europe, the EC has published a directive (European Community, 1992) aimed at protecting habitats and species in all its member states and is asking all its governments to produce a range of conservation plans and protected areas for a wide range of habitats and flora and fauna. In North America, a simple, but very important new concept that is being considered in various countries in one form or another – and has already been accepted by Canada (Department of Fisheries and Oceans, 1986) in relation to fish habitat – is that of 'no net habitat loss'. This looks to the future in relation to developments of any kind that may threaten fish habitats and makes it clear that approval will be given only if:

- there is no loss of fish habitat involved by the proposals;
- the development is modified in such a way that there is no loss of fish habitat; or
- any fish habitat that must be lost because of the development is compensated by the restoration or creation of equivalent fish habitat elsewhere in the same system, so that there is no net loss of habitat.

One of the main aims of management should be to promote diversity of habitats, which permits mobility of species, particularly where there is continuity, and thus favours species diversity.

The voluntary organizations, too, are active in promoting conservation issues – many of them relevant to fresh waters. For example, in the United Kingdom, a consortium of NGOs has produced *Biodiversity Challenge: An Agenda for Conservation Action in the UK* (Wynne *et al.*, 1995), which is forcing the government to live up to its promises at the Rio Conference and produce Species Action Plans and other programmes for conservation in the UK. In Scotland, WWF has recently launched a 'Wild Rivers' initiative (Gilvear *et al.*, 1995), which aims to move the management of rivers back towards a more natural, sustainable and (very important) more economic strategy.

These schemes and many smaller ones developing in some parts of the world are based largely on economic reasoning. If planned and implemented intelligently, they may lead to improved amenities – both recreational and aesthetic – especially in the long term. The end result is a rational use of freshwater resources leading to improved standards of living, clean lakes and rivers, appropriate recreational facilities and increased habitat protection for wildlife (Figure 10.4).

In countries where heavy population pressures have destroyed or degraded most fresh waters, this will often involve restoration or recreation of these habitats. It is therefore important that further research is promoted to refine the techniques required.

Against all the many problems facing the freshwater ecosystems of the world – lakes, rivers and wetlands – the best long-term security is integrated catchment management which gives detailed consideration to all forms of water use and needs within a river basin. Hydrological and ecological data must be incorporated into such a process, which should remain flexible and ensure that at least the more important freshwater sites remain as functioning units in the landscape. The greatest hope for fresh waters, as for other habitats, is that mankind greatly reduces both aerial and aquatic pollution to give the ecosystems of the planet conditions for life to exist.

Figure 10.4 Loch Lomond. There are numerous pressures and demands on this one important body of water. It is hoped that future management plans for the loch and its catchment will secure a sustainable future for Loch Lomond and its wildlife.

Appendix A: Glossary

Abstraction. Removal of water from an aquatic system or its catchment.
Acidification. The process (often due to human activity) by which waters become much more acid.
Algae. A varied group of simple aquatic plants, many microscopic.
Alien. An organism that is not native to the country (or geographical area) concerned.
Allochthonous. Describes material that comes in from outside the freshwater habitat.
Anuran. An amphibian in which the adults are without tails, such as frogs.
Aquifer. Water-bearing strata such as sandstone.
Autochthonous. Describes material that has been produced within the freshwater habitat.
Benthic. Dwelling on the bottom of a lake or river.
Biodiversity. The natural variety of species in a habitat.
Biosphere. That part of the earth and its atmosphere that supports living organisms.
Biota. The community of organisms present in an ecosystem.
Carr. A wet woodland area, with trees such as alder or willow, which is a successional stage between open water and dry land.
Caudata. Amphibia in which the adults have fully developed tails, such as salamanders.
Channelization. The artificial widening, deepening or straightening of a natural channel.
Dystrophic. Describes brown, peat-stained waters, which are usually acidic.
Ecosystem. A biological community together with the physical and chemical environment with which it interacts.
Ecotone. The zone of transition between adjacent ecological systems.
Epilimnion. The surface layer of warmer water that separates from the colder layer below when a lake stratifies.
Eutrophic. Rich in basic nutrients.
Eutrophication. Enrichment of fresh waters with nutrients.
Geochemistry. Chemical conditions relating to underlying rocks.
Groyne. A structure built into a bank to deflect current and prevent erosion.
Herbaceous. Describes low-growing plants that die back each winter.
Hydric soils. Soils containing a high proportion of water.

Hydrology. The study of the physical aspects of water.
Hydrophytes. Aquatic plants.
Hypolimnion. The lower layer of colder water that separates from the warmer water above when a lake stratifies.
Indigenous. Native or occurring naturally in a particular geographical area.
Lacustrine. Pertaining to a lake.
Lentic. Almost still, such as waters in lakes or ponds or very slow-flowing rivers.
Limnology. The study of inland waters.
Littoral. The edge zone along the shore of a lake or river.
Lotic. Describes flowing water as is found in streams and rivers.
Macroinvertebrates. Larger invertebrates that can easily be seen with the naked eye.
Macrophytes. Plants (mostly higher plants and bryophytes) that are easily seen with the naked eye.
Mesotrophic. Intermediate in type between nutrient-rich eutrophic and nutrient-poor oligotrophic waters.
Oligomictic. Describes lakes that have high surface temperatures, which are usually permanently stratified.
Oligotrophic. Poor in basic nutrients.
Ombrotrophic. Describes peatlands in which direct rainfall is the only supply of water.
Ox-bow lake. An ancient bend of a river which is now cut off from the main channel and contains standing water.
Palaeontology. The study of fossils.
Pelagic. Found in the open waters of lakes, away from the shore and the lake bottom.
Periphyton. Small plants (mostly algae) that are attached to solid substrates such as plants or stones.
Phytoplankton. Small plants that are suspended in the water and subject to its movements.
Restoration. Return of an ecosystem to a close approximation of its condition prior to disturbance.
Riffle. An area of fast-flowing, shallow water with a distinctly broken or disturbed surface.
Riparian. Associated with the bank of a river or lake.
Saturnism. Lead poisoning.
Soligenous. Describes peatlands in which water enters as ground water or runoff from adjacent ground, as well as from rainfall.
Stenothermic. Cold-loving.
Stratification. The separation into layers of different temperature that can take place when a lake warms up.
Stratigraphic. Concerning the succession of geological strata.
Substrate. The material making up the bed of a river or lake.

Thermocline. The layer of water (with a steep temperature gradient) in a stratified lake that lies between the upper, warmer epilimnion and the lower, cooler hypolimnion.

Water table. The level below which the ground is saturated with water.

Winterbourne. An intermittent stream.

Zooplankton. Small animals that are suspended in the water and subject to its movements.

References

Ahlgren, C. E. (1963) Some basic ecological factors in prescribed burning in North-eastern Minnesota. *Proceedings of the Annual Tall Timbers Fire Ecology Conference,* **2**, 143–149.

Almer, B., Dickson, W., Ekstrom, C. *et al.* (1974) Effects of acidification on Swedish lakes. *Ambio,* **3**, 330–336.

Altaba, C. R. (1990) The last-known population of the freshwater mussel *Margaritifera auricularia* (Bivalvia, Unionidae): a conservation priority. *Biological Conservation,* **52**, 271–286.

Andersson, C., Berggren, H. and Hamrin, S. (1975) Lake Trummen restoration project. *Verhandlungen der Internationale Vereinigung für theoretische und angewandte Limnologie,* **19**, 1097–1106.

Andrews, C. (1992) The role of zoos and aquaria in the conservation of the fishes from Africa's great lakes. *Mittelungen der Internationale Vereinigung für Limnologie,* **23**, 117–120.

Anon. (1969) Future of Lake Baikal. *Nature (London),* **223**, 1091.

Anon. (1971) *Proceedings of the International Conference of the Conservation of Wetlands and Waterfowl*, Ramsar, Iran.

Anon. (1974) Criteria and Guidelines for the choice and establishment of Biosphere Reserves. *Man and the Biosphere Report* No. 22, UNESCO, Paris, pp. 1–66.

Anon. (1976) *Proceedings of the International Conference on the Conservation of Wetlands and Waterfowl, 1974*, Heiligenhafen, Germany.

Anon. (1982) Control of vegetation at Rieselfelder, Munster, Federal Republic of Germany, in *Managing Wetlands and Their Birds,* (ed. D. A. Scott), IWRB, Gloucester, pp. 34–43.

Anon. (1990) *A Directory of Wetlands of International Importance,* IUCN, Gland.

Anon. (1991) *Acidification and Liming of Swedish Fresh Waters,* Swedish Protection Agency, Stockholm.

Anon. (1993) Controversial water project causes storm in the Okavango Delta. *Owls,* **2**, 10–12.

Anon. (1996) Convention sur les zones humides, *Le Bulletin de Ramsar No. 23*, Gland.

Armitage, P. D. (1995) Faunal community change in response to flow manipulation, in *The Ecological Basis for River Management,* (eds D. M. Harper and A. J. D. Ferguson), John Wiley, Chichester, pp. 59–91.

Armour, C. L. and Taylor, J. G. (1991) Evaluation of the instream flow incremental methodology by US Fish and Wildlife Service field users. *Fisheries,* **16**, 36–43.

Austin, M. P. and Margules, C. R. (1986) Assessing representativeness, in *Wildlife Conservation Evaluation,* (ed. M. B. Usher), Chapman & Hall, London, pp. 45–67.

Bakker, J. P. (1978) Changes in a salt marsh vegetation as a result of grazing and mowing, a five year study of permanent plots. *Vegetation,* **38**, 77–87.

Barel, C. D. N., Dorit, R., Greenwood, P. H. *et al. (1985) Destruction of fisheries in Africa's lakes. Nature* (London), **315**, 19–20.

Barendrest, A., Wassen, M. J. and Schot, P. P. (1995) Hydrological systems beyond a nature reserve; the major problem in wetland conservation of Naardermeer (The Netherlands). *Biological Conservation,* **72**, 393–405.

Battarbee, R. W. (1984) Diatom analysis and the acidification of lakes. *Philosphical Transactions of the Royal Society of London,* **305**, 193–219.

Bauman, P. C., Kitchell, J. F. and Magnuson, J. J. (1974) Lake Wingra 1837–1973. A case history of human impact. *Wisconsin Academy of Science, Arts and Letters,* **62**, 57–94.

Baxter, R. M. and Glaude, P. (1980) Environmental effects of dams and impoundments in Canada. Experience and prospects. *Canadian Bulletin of Fisheries and Aquatic Sciences,* **205**, 1–40.

Beddington, J. R. and Rettig, R. B. (1984) *Approaches to the Regulation of Fishing Effort,* Food and Agricultural Organization of the United Nations, Rome.

Belbin, L. (1993) Environmental representativeness: regional partitioning and reserve selection. *Biological Conservation,* **66**, 223–230.

Bellrose, F. C. (1959) Lead poisoning as a mortality factor in waterfowl populations. *Illinois Natural History Survey Bulletin,* **27**, 235–288.

Ben-Tuvia, A. (1981) Man-induced changes in the freshwater fish fauna of Israel. *Fisheries Management,* **12**, 139–145.

Berg, K. (1948) Biological studies on the River Susaa. *Folia Limnologica Scandinavica,* **4**, 1–318.

Berger, J. J. (1992) The Blanco River, in *Restoration of Aquatic Ecosystems,* (National Research Council), National Academy Press, Washington, pp. 470–477.

Beverton, R. J. H. and Holt, S. J. (1957) *On the Dynamics of Exploited Fish Populations,* Ministry of Agriculture, Fisheries and Food, London.

Bientema, A. J. (1982) Meadow birds in the Netherlands, in *Managing Wetlands and Their Birds,* (ed. D. A. Scott), IWRB, Gloucester, pp. 83–91.

Bishop. R. A., Andrews, R. D. and Bridges, R. J. (1979) Marsh management and its relationship to vegetation, waterfowl and musk rats. *Proceedings of the Iowa Academy of Science,* **86**, 50–56.

Björk, S. (1972) Swedish Lake Restoration Program get results. *Ambio,* **1**, 154–166.

Björk, S. (1976) The restoration of degraded wetlands, in *Proceedings of the International Conference of the Conservation of Wetlands and Waterfowl, 1974, Heiligenhafen, Germany,* (ed. M. Smart), IWRB, Gloucester, pp. 349–354.

Björk, S. and Digerfeldt, G. (1991) Development and degradation, redevelopment and preservation of Jamaican wetlands. *Ambio,* **20**, 276–284.

Blyth, J. D. (1983) Rapid stream survey to assess conservation value and habitats available for invertebrates, in *Proceedings of a Workshop on Survey Methods for Nature Conservation,* (eds K. Myers, C. Margules and I. Mustoe), CSIRO, Adelaide, South Australia.

Boon, P. J. (1991) The role of Sites of Special Scientific Interest (SSSIs) in the conservation of British rivers. *Freshwater Forum,* **1**, 95–108.

Boon, P. J., Holmes, N. T. H., Maitland, P. S. and Rowell, T. A. (1994) A system for evaluating rivers for conservation ('SERCON'): an outline of the underlying principles. *Verhandlungen der Internationalen Vereinigung für theoretische und angewandte Limnologie,* **25**, 1510–1514.

Boon, P. J., Holmes, N. T. H., Maitland, P. S. and Rowell, T. A. (1996) System for evaluating rivers for conservation. Version 1 Manual. *Scottish Natural Heritage Research, Survey and Monitoring Report No. 61*, SNH, Edinburgh.

Brakhage, G. K. (1965) Biology and behaviour of tub-nesting Canada geese. *Journal of Wildlife Management,* **29**, 751–771.

Brown, D. J. A., Howells, G. D., Dalziel, T. R. K. and Stewart, B. R. (1988) Loch Fleet. A research watershed liming project. *Water, Air and Soil Pollution,* **41**, 25–42.

Brunig, E. F. (1975) Tropical ecosystems: state and targets of research into the ecology of humid tropical systems. *Plant Research and Development,* **1**, 22–38.

Buchan, S. (1963) Conservation by integrated use of surface and ground water, in *Proceedings of a Symposium of the Institution of Civil Engineers, London, 1963,* Institution of Civil Engineers, London, pp. 181–185.

Bungenberg de Jong, C. M. (1968) Drie jaar bedrijfsonderzoek en praktijktoepassing van het herbicide diuron in de pootvisteelt. *Jaarverslag Organisatie ter Verbetering van de Binnen Visserij,* 1967–1968.

Burgis, M. J. and Symoens, J. J. (1987) *African Wetlands and Shallow Water Bodies,* ORSTOM, Paris.

Burns, J. C., Coy, J. S., Tervet, D. J. *et al.* (1984) The Loch Dee Project: a study of the ecological effects of acid precipitation and forest management on an upland catchment in south-west Scotland. 1. Preliminary investigations. *Fisheries Management,* **15**, 145–167.

Campbell, R. N., Maitland, P. S. and Campbell, R. N. B. (1994) Management of fish populations, in *The Fresh Waters of Scotland*, (eds P. S. Maitland, P. J. Boon and D. S. McLusky), John Wiley, Chichester, pp. 489–513.

Carp, E. (1980) *A Directory of Western Palaearctic Wetlands,* IUCN–UNEP, Gland.

Carpenter, K. E. (1927) Faunistic ecology of some Cardiganshire streams. *Journal of Ecology,* **15**, 33–54.

Carpenter, K. E. (1928) *Life in Inland Waters*, Sidgwick & Jackson, London.

CCS (1986) *Lochshore Management,* Countryside Commission for Scotland, Perth.

Chabreck, R. H. (1968) The relationship of cattle and cattle-grazing to marsh wildlife and plants. *Proceedings of the Annual Conference of the South-east Association of Game Fish Committees,* **22**, 55–58.

Chambers, E. G. W. (1983) *Water Supply from Loch Katrine to Glasgow and Environs*, Strathclyde Regional Council, Glasgow.

Chang, W. Y. B. (1994) Management of shallow tropical lakes using integrated lake farming. *Mittelungen der Internationale Vereinigung für Limnologie,* **24**, 219–224.

Cieslik, L. J., Harberg, M. C. and McAllister, R. F. (1993) Missouri River Master Manual review and update impact assessment methodology. *US Department of the Interior, Biological Report,* **19**, 372–386.

Clark, H. S., Mendel, H. L. and Sarbello, W. (1974) Use of artificial nest shelters in eider management. *Research in the Life Sciences, University of Maine,* **22**, 1–15.

Cohen, A., Bills, R., Cocquyt, C. Z. and Caljon, A. G. (1993) The impact of sediment pollution on biodiversity in Lake Tanganyika. *Conservation Biology,* 7, 667–674.

Cooke, A. S. (ed.) (1986) *The Use of Herbicides on Nature Reserves*, Nature Conservancy Council, Peterborough.

Cooke, G. D. and Martin, A. B. (1989) Long-term evaluation of the effectiveness of phosphorus inactivation. *Annual Meeting of the North American Lake Management Society*, **1**, 1–5.

Coulter, G. W. (ed.) (1991) *Lake Tanganyika and Its Life*, Oxford University Press, Oxford.

Coulter, G. W. and Mubamba, R. (1993) Conservation in Lake Tanganyika, with special reference to underwater parks. *Conservation Biology*, **7**, 678–685.

Cowardin, L. M., Carter, V., Golet, F. C. *et al.* (1977) *Classification of Wetland and Deep-water Habitats of the United States (An Operational Draft)*, US Fish and Wildlife Service, Washington, DC.

Crivelli, A. J. and Maitland, P. S. (eds) (1995) Endemic freshwater fishes of the northern Mediterranean region. *Biological Conservation*, **72**.

Davison, W., George, G. D. and Edwards, N. J. A. (1995) Controlled reversal of lake acidification by treatment with phosphate fertiliser. *Nature (London)*, **377**, 504–507.

De Angelis, D. L. and Cushman, R. M. (1990) Potential application of models in forecasting the effects of climate changes on fisheries. *Transactions of the American Fisheries Society*, **119**, 224–239.

Department of Fisheries and Oceans (1986) *Policy for the Management of Fish Habitat*, Department of Fisheries and Oceans, Ottawa.

De Waal, L. C., Child, L. E. and Wade, M. (1995) The management of three alien invasive plants, in *The Ecological Basis for River Management*, (eds D. M. Harper and A. J. D. Ferguson), John Wiley, Chichester, pp. 315–321.

Dimentman, C., Bromley, H. J. and Por, F. D. (1992) *Lake Hula*, Israel Academy of Science and Humanities, Jerusalem.

Djoshkin, W. W. and Safonov, W. G. (1972) *Die Biber der alten und nauen Welt*, Neue Brehm Bucherei, Witlemberg-Lutherstadt.

Dolmen, D. (1987) *Gyrodactylus salaris* (Monogenea) in Norway; infestations and management, in *Proceedings of the Symposium on Parasites and Diseases in Natural Waters and Aquaculture in Nordic Countries, 1986*, (eds A. Stenmark and G. Malmberg), University of Stockholm, Stockholm, pp. 63–69.

Drinkwater, K. F. and Frank, K. T. (1994) Effects of river regulation and diversion on marine fish and invertebrates. *Aquatic Conservation*, **4**, 135–151.

Duffey, E. (1962) The Norfolk Broads. A regional study of wildlife conservation in a wetland area with high tourist attraction. *Project MAR, IUCN Publication*, **3**, 290–301.

Duffey, E. (1968) Ecological studies on the large copper butterfly at Woodwalton Fen NNR Huntingdonshire. *Journal of Applied Ecology*, **5**, 69–96.

Duffey, E. and Morris, M. G. (eds) (1965) The conservation of invertebrates. Monks Wood Experimental Station Symposium No 1, Mimeographed report.

Dugan, P. J. (1990) *Wetland Conservation*, IUCN, Gland.

Duncan, P. and d'Herbes, J. M. (1982) The use of domestic herbivores in the management of wetlands for waterbirds in the Camargue, France, in *Managing Wetlands for Birds*, (eds I. Fog, T. Lampio, J. Rooth and M. Smart), Nimsfielde Press, Gloucester, pp. 51–66.

Dykyjova, D. and Kvet, J. (1978) *Pond Littoral Ecosystems. Structure and Functioning*, Springer-Verlag, Berlin.

Edmondson, W. T. (1991) *The Uses of Ecology: Lake Washington and Beyond*, University of Washington Press, Seattle, WA.

Edwards, R. W. (1958) The effect of larvae of *Chironomus riparius* Meigen on the redox potentials of settled activated sludge. *Annals of Applied Biology*, **46**, 457–464.

Edwards, P. (1980) Food potential of aquatic macrophytes. *ICLARM Report*, 5, 1–51.

Egglesmann, R. (1987) Okotechnische Aspekte de Hochmoor Regeneration. *Telma*, **17**, 59–94.

Ellery, W. N. and McCarthy, T. S. (1994) Principles for the sustainable utilisation of the Okavango Delta ecosystem, Botswana. *Biological Conservation*, **70**, 159–168.

Elliott, J. M. and Tullett, P. A. (1992) The medicinal leech. *Biologist*, **39**, 153–158.

Ellis, E. A. (1963) Some effects of selective feeding by the Coypu, *Myocastor coypus*, on the vegetation of Broadland. *Transactions of the Norfolk and Norwich Natural History Society*, **20**, 32–35.

Eriksson, M. O. G. (1984) Acidification of lakes: effects on waterbirds in Sweden. *Ambio*, **13**, 260–262.

Etherington, J. R. (1983) *Wetland Ecology*, Edward Arnold, London.

European Community (1992) *Directive on the Conservation of Natural Habitats and Wild Fauna and Flora*, European Community, Brussels.

Ferdinand, L. (1977) The Danish system of registration and codifying bird localities, in *Proceedings of the Technical Meeting on Evaluation of Wetlands from a Conservation Point of View, Bonn*, IWRB, Gloucester, pp. 37–44.

Fiala, K. and Kvet, J. (1971) Dynamic balance between plant species in South Moravian reed swamps, in Duffey, E. and Watt, A. S. (eds) *The Scientific Management of Animal and Plant Communities for Conservation*, Blackwell, Oxford, pp. 241–269.

Fournier, O. and Wattier, J. M. (1979) *Introduction à la question des oiseaux d'eau et des zones humides. 1. Données générales*, ONC, Paris.

Fox, P. J. (1976) Preliminary observations on fish communities of the Okovanga Delta, in *Proceedings of a Symposium on the Okovanga Delta and Its Future Utilisation*, Botswana Society, Gaberone, pp. 125–130.

Frederickson, A. H. and Taylor, T. S. (1982) Management of seasonally flooded impoundments for wildlife, *Resource Publication 148, US Fish and Wildlife Service*, Washington, DC, pp. 1–22.

Frieberg, N., Kronvang, B., Svendsen, L. M. and Hansen, H. O. (1994) Restoration of a channelised reach of the River Gels, Denmark: effects on the macroinvertebrate community. *Aquatic Conservation*, **4**, 289–296.

Gagliano, S. M., Meyer-Arendt, K. J. and Wicker, K. M. (1981) Land loss in the Mississippi deltaic plain. *Transactions of the Gulf Coast Association of Geological Societies*, **31**, 295–300.

George, M. (1992) *The Land-use, Ecology and Conservation of Broadland*, Packard, Chichester.

Getz, W. M. and Haight, R. G. (1989) *Population Harvesting: Demographic Models of Fish, Forest and Animal Resources*, Princeton University Press, Princeton, NJ.

Giles, N. (1992) *Wildlife After Gravel*, The Game Conservancy, Fordingbridge.

Gilvear, D. J., Hanley, N., Maitland, P. S. and Peterken, G. (1995) *Wild Rivers. Phase 1: Technical Paper*, Worldwide Fund for Nature, Perth.

Glantz, M. H. (1990) Does history have a future – forecasting climate change effects on fisheries by analogy. *Fisheries*, **15**, 39–44.

Goldsmidt, T., Witte, F. and Wanink, J. (1993) Cascading effects of the introduced Nile perch on the detritivorous/planktivorous species in the sublittoral areas of Lake Victoria. *Conservation Biology*, 7, 686–700.

Goldsmith, F. B. (1983) Evaluating nature, in *Conservation in Perspective*, (eds A. Warren and F. B. Goldsmith), John Wiley, Chichester, pp. 233–246.

Goldsmith, F. B. (ed.) (1994) *Monitoring for Conservation and Ecology*, Chapman & Hall, London.

Goode, D. (1977) Peatlands, in *A Nature Conservation Review*, (ed. D. A. Ratcliffe), Cambridge University Press, Cambridge, **1**, pp. 249–287; **2**, 206–244.

Goodfellow, S. and Peterken, G. F. (1981) A method for survey and assessment of woodlands for nature conservation using maps and species lists: the example of Norfolk woodlands. *Biological Conservation*, **21**, 177–195.

Goodwillie, R. (1980) *European Peatlands*, Council of Europe, Strasbourg.

Gopal, B., Kvet, J., Loffler, H. *et al.* (1990) Definition and classification, in *Wetlands and Shallow Continental Water Bodies*, (ed. B. C. Patten), Academic Publishing, The Hague, pp. 9–15.

Gordon, I. J., Duncan, P., Grillas, P. and Lecomte, T. (1990) The use of domestic herbivores in the conservation of the biological richness of European wetlands. *Bulletin of Ecology*, **21**, 49–60.

Gorham, E. (1961) Factors influencing supply of major ions to inland waters, with special reference to the atmosphere. *Geological Society of America Bulletin*, **72**, 795–840.

Goulding, M. (1989) *Amazon: The Flooded Forest*, BBC, London.

Granval, P. (1988) Quantified appreciation of land parcel characteristics in a wet zone as basis of the rural valorisation. *Proceedings of the International Wetlands Conference*, **3**, 165.

Grimaldi, E. and Numann, W. (1972) The future of salmonid communities in European subalpine lakes. *Journal of the Fisheries Research Board of Canada*, **29**, 931–936.

Gryseels, M. (1989a) Nature management experiments in a derelict reedmarsh. I: Effects of winter cutting. *Biological Conservation*, **47**, 171–193.

Gryseels, M. (1989b) Nature management experiments in a derelict reedmarsh. II: Effects of summer mowing. *Biological Conservation*, **48**, 85–99.

Grzimek, B. (1965) *Wildes Tier. Weisser Mann Munchen*, Kindler Verlag, Munich.

Gulland, J. A. (1989) *Fish Stock Assessment: A Manual of Basic Methods*, John Wiley/FAO, Chichester.

Gurnell, A. M., Gregory, K. J. and Petts, G. E. (1995) The role of coarse woody debris in forest aquatic habitats: implications for management. *Aquatic Conservation*, **5**, 143–166.

Hafner, H. (1977) Contribution à l'étude écologique de quatre espèces de herons (*Egretta g. garzetta* L., *Ardeola r. ralloides* Scop., *Ardeola i. ibis* L., *Nycticorax n. nycticorax* L.) pendant leur nidification en Camargue. Unpublished thesis, Toulouse.

Hafner, H. (1982) The creation of a breeding site for tree-nesting herons in the Camargue, France, in *Managing Wetlands and Their Birds*, (ed. D. A. Scott), IWRB, Gloucester, pp. 216–220.

Haines, T. A. (1981) Acidic precipitation and its consequences for aquatic ecosystems: a review. *Transactions of the American Fisheries Society*, **110**, 669–707.

Haller, W. T, Shireman, J. V. and Durant, D. F. (1980) Fish harvest resulting from mechanical control of *Hydrilla. Transactions of the American Fisheries Society,* **109**, 517–520.

Hansen, H. O. (ed.) (1996) *River Restoration – Danish Experience and Examples,* National Environment Research Institute, Silkeborg, 99 pp.

Harrison, J. (ed.) (1974) *Caerlaverock – conservation and wildfowling in action.* WAGBI, Gloucester.

Harrison, J. G. (1972) *The Gravel Pit Waterfowl Reserve, Sevenoaks. Artificial* Raft Islands for Waterfowl. Manual of Wetland Management, IWRB, Gloucester.

Harrison, J. G. (1982) Creating and improving wading bird habitat at Sevenoaks, England, in *Managing Wetlands and Their Birds,* (ed. D. A. Scott), IWRB, Gloucester, pp. 137–142.

Harrison, A. D. and Elsworth, J. F. (1958) Hydrobiological studies on the Great Berg River, Western Cape Province. *Transactions of the Royal Society of South Africa,* **35**, 125–329.

Haslam, S. M. (1969) A study of *Phragmites communis* Trin. in relation to its cultivation and harvesting in East Anglia for the thatching industry. *Monograph No. 1,* Norfolk Reed Growers' Association, Norwich.

Haslam, S. M. (1972a) Biological flora of the British Isles. *Phragmites communis* Trin. *Journal of Ecology,* **60**, 585–610.

Haslam, S. M. (1972b) *The Reed (Norfolk Reed),* Norfolk Reed Growers' Association, Norwich.

Haslam, S. M. (1973) The management of British wetlands. II. Conservation. *Journal of Environmental Management,* **1**, 345–361.

Haslam, S. M. (1975) River vegetation and pollution, in *Science, Technology and Environmental Management,* (eds R. D. Hay and T. D. Davies), Saxon House, Lexington, KY, pp. 137–143.

Hasler, A. D. (1947) Eutrophication of lakes by domestic drainage. *Ecology,* **28**, 383–395.

Hassan, P. (1993) Aspects juridiques de la conservation des zones humides. *Ramsar Bulletin,* **15**, 11.

Healey, M. C. (1990) Implications of climate change for fisheries management policy. *Transactions of the American Fisheries Society,* **119**, 366–373.

Henderson-Sellars, B. and Markland, H. R. (1987) *Decaying Lakes. The Origins and Control of Cultural Eutrophication,* John Wiley, Chichester.

Hillebrand, D. (1950) Verkrautung und Abfluss. *Deutsche gewasserkundliche Mitteilungen,* **2**, 1–30.

Hocutt, C. H. and Stauffer, J. R. (1980) *Biological Monitoring of Fish,* Lexington Books, Lexington, KY.

Holden, A. V. (1966) A chemical study of rain and stream waters in the Scottish highlands. *Freshwater and Salmon Fisheries Research, Scotland,* **37**, 1–17.

Hopthrow, H. E. (1963) Utilisation of water in industry, in *Proceedings of a Symposium of the Institution of Civil Engineers, London, 1963,* Institution of Civil Engineers, London, pp. 30–33.

Horton, R. E. (1945) Erosional development of streams and their drainage basins. *Bulletin of the Geological Society of America,* **56**, 275–370.

Howell, D. L. (1994) Role of environmental agencies, in *The Fresh Waters of*

Scotland, (eds P. S. Maitland, P. J. Boon and D. S. McLusky), John Wiley, Chichester, pp. 577–611.

Isom, B. C. and Hudson, R. G. (1982) In vitro culture of parasitic freshwater mussel glochidia. *Nautilus*, **96**, 147–151.

IUCN, UNEP and WWF (1991) *Caring for the Earth. A Strategy for Sustainable Living*, IUCN, UNEP and WWF, Gland.

Jeffery, R. C. V., Chabwela, H. N., Howard, G. and Dugan, P. J. (eds) (1992) Managing the wetlands of Kafue Flats and Bangweula Basin. *Proceedings of the WWF Zambia Wetland Project Workshop, Masungwa Safari Lodge, Kafui National Park, Zambia, 1986*, IUCN, Gland.

Johnson, J. E. and Rinne, J. N. (1982) The Endangered Species Act and southwest fishes. *Bulletin of the American Fisheries Society*, **7**, 1–8.

Kallemeyn, L. W. (1983) Action plan for aquatic research at Voyageurs National Park. *Park Science*, **4**, 18.

Kallemeyn, L. W., Cohen, Y. and Radomski, P. (1993) Rehabilitating the aquatic ecosystem of Rainy Lake and Namakan Reservoir by restoration of a more natural hydrologic regime. *US Department of the Interior, Biological Report*, **19**, 432–448.

Kallio-Nyberg, I. and Koljonen, M. (1991) The Finnish char (*Salvelinus alpinus*) stock register. *Finnish Fisheries Research*, **12**, 77–82.

Keen, E. A. (1988) *Ownership and Productivity of Marine Fishery Resources*, McDonald & Woodward, Blacksburg, VA.

Kennedy, V. S. (1990) Anticipated effects of climate change on estuarine and coastal fisheries. *Fisheries*, **15**, 16–24.

Klein, L. (1957) *Aspects of River Pollution*, Butterworths, London.

Koljonen, M. and Kallio-Nyberg, I. (1991) The Finnish trout (*Salmo trutta*) stock register. *Finnish Fisheries Research*, **12**, 83–90.

Komarek, E. V. (1965) Fire ecology – grasslands and man. *Proceedings of the Annual Tall Timbers Fire Ecology Conference*, **4**, 169–220.

Laan, R. and Verboom, B. (1990) Effect of pool size and isolation on amphibian communities. *Biological Conservation*, **54**, 251–262.

Lagler, K. F. (1949) *Studies in Freshwater Fishery Biology*, University of Michigan, Ann Arbor, MI.

Lake District National Park (1993) Bassenthwaite Lake management plan. Lake District Special Planning Board, Kendal.

Lamberson, R. H., McKelvey, R., Noon, B. R. and Voss, C. (1992) A dynamic analysis of northern spotted owl viability in a fragmented forest landscape. *Conservation Biology*, **6**, 505–512.

Larkin, P. A. (1977) An epitaph for the concept of MSY. *Transactions of the American Fisheries Society*, **107**, 1–11.

Larsson, T. (1982) Restoration of lakes and other wetlands in Sweden, in *Managing Wetlands and Their Birds*, (ed. D. A. Scott), IWRB, Gloucester, pp. 107–122.

Lecomte, T. and Leneveu, C. (1986) Le Marais Vernier: Contribution à l'étude et à la question d'une zone humide. University of Rouen, Thesis.

Le Cren, E. D. (1990) Rare fishes and their conservation: a brief introduction to the symposium. *Journal of Fish Biology*, **37A**, 1–3.

Levieil, D., Cutipa, Q. C., Goyzueta, C. G. and Paz, F. P. (1989) The socio-economic importance of macrophyte extraction in Puno Bay, in *Pollution in Lake*

Titicaca, (eds T. G. Northcote, P. Morales, D. A. Levy and M. S. Greaven), University of British Columbia, Vancouver, p. 262.

Lillehammer, A. and Saltveit, S. J. (eds). (1984) *Regulated Rivers*, Universitetsforlaget As, Oslo.

Linde, A. F. (1969) Techniques for wetland management. *Research Report*, 45, Department of Natural Resources, Madison, WI, pp. 1–156.

Loch Lomond and Trossachs Working Party (1993) *The Management of Loch Lomond and the Trossachs*. Scottish Office, Edinburgh.

Löffler, H. (1990) Human uses, in *Wetlands and Shallow Continental Water Bodies*, (ed. B. C. Patten), Academic Publishing, The Hague, pp. 759.

Lomolino, M. V. (1994) An evaluation of alternative strategies for building networks of nature reserves. *Biological Conservation*, **69**, 243–249.

Lumsden, H. G. (1982) Artificial nesting structures for birds, in *Managing Wetlands and Their Birds*, (ed. D. A. Scott), IWRB, Gloucester, pp. 179–199.

Lund, J. W. G. (1950) Studies on *Asterionella formosa* Hass. II. Nutrient depletion and the spring maximum. *Journal of Ecology*, **38**, 1–35.

Luther, H. and Rzoska, J. (1971) *Project Aqua. A Source Book of Inland Waters Proposed for Conservation*, Blackwell, Oxford.

Macdonald, L. H. (1991) *Monitoring Guidelines to Evaluate Effects of Forestry Activities on Streams in the Pacific Northwest and Alaska*, US Experimental Protection Agency, Seattle, WA.

Madsen, B. L. (1995) Danish watercourses – ten years with the new Watercourse Act. *Miljonyt*, **11**, 209.

Maitland, P. S. (1966) *The Fauna of the River Endrick*, Blackie, Glasgow.

Maitland, P. S. (1972) *A Key to the Freshwater Fishes of the British Isles, With Notes on Their Distribution and Ecology*, Freshwater Biological Association, Ambleside.

Maitland, P. S. (1979) *Synoptic Limnology: The Analysis of British Freshwater Ecosystems*, Institute of Terrestrial Ecology, Cambridge.

Maitland, P. S. (1984) The effects of eutrophication on wildlife. *Institute of Terrestrial Ecology Symposium*, **13**, 101–108.

Maitland, P. S. (1985a) Criteria for the selection of important sites for freshwater fish in the British Isles. *Biological Conservation*, **31**, 335–353.

Maitland, P. S. (1985b) The status of the River Dee in a national and international context. *Institute of Terrestrial Ecology Symposium*, **14**, 142–148.

Maitland, P. S. (1986) *Conservation of Threatened Freshwater Fish in Europe*, Council of Europe, Strasbourg.

Maitland, P. S. (1987a) Fish in the Clyde and Leven systems – a changing scenario. *Proceedings of the Institute of Fisheries Management Conference, 1987*, Institute of Fisheries Management, Pitlochry, pp. 13–20.

Maitland, P. S. (1987b) Conserving fish in Australia. *Proceedings of the Conference on Australian Threatened Fish, Melbourne, 1985*, Australian Society for Fish Biology, Melbourne, Victoria, pp. 63–67.

Maitland, P. S. (1987c) Fish conservation: a world strategy. *Annual Bulletin of the Freshwater Fish Protection Association, Japan*, 10–21.

Maitland, P. S. (1990) *The Biology of Fresh Waters*, 2nd edn, Blackie, Glasgow.

Maitland, P. S. (1991) Climate change and fish in northern Europe: some possible scenarios. *Proceedings of the Institute of Fisheries Management, Annual Study Course*, **22**, 97–110.

Maitland, P. S. (1992a) Fish conservation in Europe: wise exploitation of a valuable resource. *Proceedings of a Council of Europe Meeting, Milan.*

Maitland, P. S. (1992b) Fish conservation in the British Isles: the role of captive breeding. *Animals Magazine,* **2**, 25–28.

Maitland, P. S. (1994) Fish, in *The Fresh Waters of Scotland,* (eds P. S. Maitland, P. J. Boon and D. S. McLusky), John Wiley, Chichester, pp. 191–208.

Maitland, P. S. and Evans, D. (1986) The role of captive breeding in the conservation of fish species. *International Zoo Yearbook,* **25**, 66–74.

Maitland, P. S. and Lyle, A. A. (1990) Practical conservation of British fishes: current action on six declining species. *Journal of Fish Biology,* **37A**, 255–256.

Maitland, P. S. and Lyle, A. A. (1991) Conservation of freshwater fish in the British Isles: the status and biology of threatened species. *Aquatic Conservation,* **1**, 25–54.

Maitland, P. S. and Lyle, A. A. (1992) Conservation of freshwater fish in the British Isles: proposals for management. *Aquatic Conservation,* **2**, 165–183.

Maitland, P. S., East, K. and Morris K. H. (1983) Ruffe *Gymnocephalus cernua*, new to Scotland, in Loch Lomond. *Scottish Naturalist,* **1983**.

Maitland, P. S., Lyle, A. A. and Campbell, R. N. B. (1987) *Acidification and Fish Populations in Scottish Lochs,* Institute of Terrestrial Ecology, Grange-over-Sands.

Maitland, P. S., Newson, M. D. and Best, G. E. (1990) *The Impact of Afforestation and Forestry Practice on Freshwater Habitats*, Nature Conservancy Council, Peterborough.

Maitland, P. S., May, L., Jones, D. H. and Doughty, C. R. (1991) Ecology and conservation of Arctic Charr, *Salvelinus alpinus* (L.), in Loch Doon, an acidifying loch in southwest Scotland. *Biological Conservation,* **55**, 167–197.

Margules, C. R. and Usher, M. (1984) Conservation evaluation in practice: 1. Sites of different habitats in north-east Yorkshire, Great Britain. *Journal of Environmental Management,* **18**, 153–168.

Margules, C. R., Nicholls, A. O. and Pressey, R. L. (1988) Selecting networks of reserves to maximise biological diversity. *Biological Conservation,* **43**, 63–76.

Marsh, C. M. (1963) Use of water for navigation, in *Proceedings of a Symposium of the Institution of Civil Engineers, London, 1963*, Institution of Civil Engineers, London, pp. 52–58.

Martin, A., Erickson, R. C. and Stennis, J. H. (1957) Improving duck marshes by weed control. *Circular No. 19*, US Fish and Wildlife Service, Washington, DC, pp. 1–60.

Mason, C. F. and Madsen, A. B. (1990) Mortality and condition in otters *Lutra lutra* from Denmark and Great Britain. *Natura Jutlandica,* **24**, 217–220.

Meade, R. (1992) Some early changes following the wetting of a vegetated cutover peatland surface at Danes Moss, Cheshire, UK and their relevance to conservation management. *Biological Conservation,* **61**, 31–40.

Meier, T. I. (1981) Artificial nesting structures for the Double-crested Cormorant. *Wisconsin Department of Natural Resources, Technical Bulletin,* **126**, 1–12.

Mikola, J., Miettinen, M., Lehikoinen, E. and Lehtila, K. (1994) The effects of disturbance caused by boating on survival and behaviour of velvet scoter *Melanitta fusca* ducklings. *Biological Conservation,* **67**, 119–124.

Moore, N. W. (1964) Intra and interspecific competition among dragonflies (Odonata). *Journal of Animal Ecology,* **33**, 49–71.

Moore, N. W. (1991) The development of dragonfly communities and the consequences of territorial behaviour; a 27-year study on small ponds at Woodwalton Fen, Cambridgeshire, United Kingdom. *Odonatologica,* **20**, 203–231.

Morgan, N. C. (1970) Changes in the flora and fauna of a nutrient enriched lake. *Hydrobiologia,* **35**, 543–553.

Morgan, N. C. (1972) Problems of the conservation of freshwater ecosystems. *Symposium of the Zoological Society of London,* **29**, 135–154.

Morgan, N. C. (ed.) (1975) *Report of the Technical Working Party – Cotswold Water Park Aquatic Nature Reserve*, Nature Conservancy Council, Peterborough.

Morgan, N. C. (1978) Towards improved criteria for selecting wetlands for wildfowl conservation, in *Technical Meeting of Evaluation of Wetlands from a Conservation Point of View, Bonn, 8–9 October 1977*, IWRB, Gloucester, pp. 15–19.

Morgan, N. C. (1982) An ecological survey of standing waters in North West Africa: II. Site descriptions for Tunisia and Algeria. *Biological Conservation,* **24**, 83–113.

Morgan, N. C. (1990) Nature reserves, in *Wetlands and Shallow Continental Water Bodies*, (ed. B. C. Patten), Academic Publishing, The Hague, pp. 603–622.

Morgan, N. C. and Boy, V. (1982) An ecological survey of standing waters in North West Africa: 1. Rapid survey and classification. *Biological Conservation,* **24**, 5–44.

Morgan, N. C. and Britton, R. H. (1977) Open waters, in *A Nature Conservation Review*, (ed. D. A. Ratcliffe), Cambridge University Press, Cambridge, **1**, pp. 201–248; **2**, 166–205.

Mudge, G. P. (1983) The incidence and significance of ingested lead pellet poisoning in British waterfowl. *Biological Conservation,* **27**, 333–372.

Murphy, K. J. and Pearce, H. G. (1987) Habitat modification associated with freshwater angling. *Institute of Terrestrial Ecology Symposium,* **19**, 31–46.

Murray, J. (1910) Characteristics of lakes in general and their distribution over the surface of the globe, in *Bathymetrical Survey of the Freshwater Lochs of Scotland,* (eds J. Murray and L. Pullar), Challenger, Edinburgh.

Musil, O. (1973) Technical interventions in the Lednice fishponds national nature reserve, in *Littoral of the Nesyt Fishpond*, (ed. J. Kvet), Academia, Prague, pp. 157–158.

National Rivers Authority (1991) *Catchment Management Planning*, NRA, Bristol.

Nature Conservancy Council (1989) *Guidelines for Selection of Biological SSSIs,* Nature Conservancy Council, Peterborough.

Newall, A. M. (1995) The microflow environments of aquatic plants – an ecological perspective, in *The Ecological Basis for River Management,* (eds D. M. Harper and A. J. D. Ferguson), John Wiley, Chichester, pp. 79–92.

Newbold, C. (1975) Herbicides in aquatic systems. *Biological Conservation,* 7, 97–118.

Newbold, C. (1984) Aquatic and bankside herbicides, in *Rivers and Wildlife Handbook,* (eds G. Lewis and G. Williams), RSPB/RSNC, Sandy, pp. 209–214.

Newbold, C., Purseglove, J. and Holmes, N. (1983) *Nature Conservation and River Engineering,* Nature Conservancy Council, Peterborough.

Newman, J. F. (1967) The ecological effects of bypridyl herbicides used for aquatic weed control. *Proceedings of the European Weed Research Council International Symposium on Aquatic Weeds,* **2**, 1–13.

Nick, K. J. (1984) Measures and chances of success for the regeneration of bogs after the complete industrial cutting of peat. *Proceedings of the International Peat Congress*, 7, 331–338.

Nilsson, C. (1981) Dynamics of the shore vegetation of a North Swedish hydro-electric reservoir during a 5-year period. *Acta phytogeographica suedica*, **69**, 1–94.

O'Keefe, J. H., Danilewitz, D. B. and Bradshaw, J. A. (1987) An 'expert system' approach to the assessment of the conservation status of rivers. *Biological Conservation*, **40**, 69–84.

Oleinikov, N. S. (1971) *Artificial Breeding Sites of Wild Ducks*, US Dept of the Interior and National Science Foundation Translations, Indian Scientific Documentation Centre, New Delhi.

Olsen, S. (1964) Vegetation saendringer i Lingby Sø. Bidrag til analyse af kulturpåvirkninger på vand- og sumpplante vegetationen. *Botaniske Tidsskriften*, **59**, 273–300.

Owen, M. (1982) Management of summer grazing and winter disturbance on goose pasture at Slimbridge, England, in *Managing Wetlands and Their Birds*, (ed. D. A. Scott), IWRB, Gloucester, pp. 67–72.

Pakarinen, P. (1994) Impacts of drainage on Finnish peatlands and their vegetation. *International Journal of Ecology and Environmental Sciences*, **20**, 173–183.

Pearsall, W. H. (1921) The aquatic vegetation of the English lakes. *Journal of Ecology*, **8**, 163–201.

Pearson, R. G. and Jones, N. V. (1975) The effects of dredging operations on the benthic community of a chalk stream. *Biological Conservation*, **8**, 273–278.

Petts, G., Moller, H. and Roux, A. L. (eds) (1989) *Historical Changes of Large Alluvial Rivers: Western Europe*, John Wiley, Chichester.

Pianka, E. R. (1966) Latitudinal gradients in species diversity. *American Naturalist*, **100**, 33–46.

Pierce, N. D. (1970) Inland lake dredging evaluation. *Department of Natural Resources, Madison, Wisconsin. Technical Bulletin*, **46**, 1–68.

Pierce, G. J., Spray, C. J. and Stuart, E. (1993) The effect of fishing on the distribution and behaviour of waterbirds in the Kukut area of Lake Songkla, southern Thailand. *Biological Conservation*, **66**, 23–34.

Pieterse, A. H. and Murphy, K. J. (eds) (1990) *Aquatic Weeds*, Oxford University Press, Oxford.

Pitcher, T. J. and Hart, P. J. B. (1982) *Fisheries Ecology*, Croom Helm, London.

Pollard, E. and Yates, T. J. (eds) (1995) *Monitoring Butterflies for Ecology and Conservation*, Chapman & Hall, London.

Pollard, D. A., Ingram, B. A., Harris, J. H. and Reynolds, L. F. (1990) Threatened fishes in Australia – an overview. *Journal of Fish Biology*, **37A**, 67–78.

Postupalsky, S. (1978) Artificial nesting sites for Ospreys and Bald Eagles, in *Endangered Birds – Management Techniques for Preserving Threatened Species*, (ed. S. A. Temple), Croom Helm, Madison, WI, pp. 35–45.

Pretscher, P. (1983) *Hinweise zur Gestaltung eines Libellengewassers. Schutzprogramm für Tagfalter und Widderchen in Hamburg*, Hamburg.

Price, H. (1981) A review of current mechanical methods. *Proceedings of the AAB Conference on Aquatic Weeds and their Control*, **1981**, 77–86.

Prickett, C. N. (1963) Use of water for agriculture, in *Proceedings of a Symposium of the Institution of Civil Engineers, London, 1963*, Institution of Civil Engineers, London, pp. 15–29.

Probst, A. Massabau, J. C., Probst, J. L. and Bertrand, F. (1990) Acidification des eaux de surface sous l'influence des précipitations acides: rôle de la végétation et du substratum, consequences pour les populations de truites. Le cas des ruisseaux des Vosges. *Compte rendu de l'Académie de Sciences, Paris,* **311**, 405–411.

Rabe, F. W. and Gibson, F. (1984) The effect of macrophyte removal on the distribution of selected invertebrates in a littoral environment. *Journal of Freshwater Ecology,* **2**, 359–371.

Rabe, F. W. and Savage, N. L. (1979) A methodology for the selection of aquatic natural areas. *Biological Conservation,* **15**, 291–300.

Ratcliffe, D. A. (ed.) (1977) *A Nature Conservation Review,* Cambridge University Press, Cambridge.

Rees, W. J. (1965) The aerial dispersal of Mollusca. *Proceedings of the Malacological Society, London,* **36**, 269–282.

Regier, H. A. and Meisner, J. D. (1990) Anticipated effects of climate change on freshwater fishes and their habitat. *Fisheries,* **15**, 10–15.

Reichholf, J. F. (1977) On the evaluation of a wetland from an ecological point of view, in *Proceedings of Technical Meeting of Evaluation of Wetlands from a Conservation Point of View,* IWRB, Bonn, pp. 11–13.

Reid, G. M. (1990) Captive breeding for the conservation of cichlid fishes. *Journal of Fish Biology,* **37A**, 157–166.

Reinthal, P. (1993) Evaluating biodiversity and conserving Lake Malawi's cichlid fish fauna. *Conservation Biology,* 7, 712–718.

Rennie, P. J. (1957) The uptake of nutrients by timber forest and its importance to timber forest in Britain. *Quarterly Journal of Forestry,* **51**, 101–105.

Ricker, W. E. (1934) An ecological classification of Ontario streams. *Publications of the Ontario Fisheries Research Laboratory,* **49**, 1–114.

Ricker, W. E. (1954) Stock and recruitment. *Journal of the Fisheries Research Board of Canada,* **11**, 559–623.

Ruddle, K. (1980) A preliminary survey of fish culture in rice fields, with special reference to West Java, Indonesia. *Bulletin of the National Museum of Ethnology,* **5**, 801–822.

Ryder, R. A. and Kerr, S. R. (1984) Reducing the risk of fish introductions: a rational approach to the management of cold-water communities. *EIFAC Technical Paper,* **42**, 510–533.

Rzoska, J. (1976) A controversy reviewed – the arguments over the Aswan High Dam on the River Nile continue. *Nature (London),* **261**, 444–445.

Saetersdal, M. and Birks, H. J. B. (1993) Addressing the representativeness of nature reserves using multivariate analysis: vascular plants and breeding birds in deciduous forests, western Norway. *Biological Conservation,* **65**, 121–132.

Saetersdal, M., Line, J. M. and Birks, H. J. B. (1993) How to maximise biological diversity in nature reserve selection: vascular plants and breeding birds in deciduous woodlands, western Norway. *Biological Conservation,* **66**, 131–138.

Samways, M. J. (1989) Farm dams as nature reserves for dragonflies (Odonata) at various altitudes in the Natal Drakensberg Mountains, South Africa. *Biological Conservation,* **48**, 181–187.

Sanchez-Perez, J. M., Tremolieres, M. and Carbonier, R. (1991) Une station d'épuration naturelle des phosphates et nitrates apportés par des eaux de débordement du Rhin: la forêt alluviale à frêne et orme. *Compte rendu de l'Académie de Sciences, Paris,* **312**, 395–402.

Sandilands, A. P. (1980) Artificial nesting structures for Great Blue Herons. *Blue Jay,* **38**, 187–188.

Sawyer, F. (1985) *Keeper of the Stream – The Life of a River and Its Trout Fishery,* Allen & Unwin, London.

Schiemer, F. and Waidbacher, H. (1992) Strategies for conservation of a Danubian fish fauna, in *River Conservation and Management,* (eds P. J. Boon, P. Calow and G. Petts), John Wiley, Chichester, pp. 363–382.

Schiemer, F., Spindler, T. and Wintersberger, H. (1991) Fish fry associations: important indicators for the ecological status of large rivers. *Verhandlungen der Internationale Vereinigung für theoretische und angewandte Limnologie,* **24**, 2497–2500.

Schlupp, I. and Podloncky, R. (1994) Changes in breeding site fidelity: a combined study of conservation and behaviour in the common toad, *Bufo bufo*. *Biological Conservation,* **69**, 285–291.

Schmitz, W. (1955) Physiographische Aspekte der limnologischen Fliesgewasserntypen. *Archiv für Hydrobiologie,* **22**, 510–523.

Schogolev, I. (1996) Wetland ecosystems of the north coast of the Black Sea: their destruction by human activities. *Biological Conservation,* in press.

Schrader-Frechette, K. S. and McCoy, E. D. (1994) *Method in Ecology*, Cambridge University Press, Cambridge.

Scott, A. (1979) Development of economic theory on fisheries regulation. *Journal of the Fisheries Research Board of Canada,* **36**, 725–741.

Scott, D. A. (ed.) (1989) *A Directory of Asian Wetlands*, IUCN, Gland.

Scott, D. A. and Carbonell, M. (1986) *A Directory of Neotropical Wetlands*, IUCN, Gland.

Scudder, T., Manley, R. E., Coley, R. W. *et al.* (1993) *The IUCN Review of the Southern Okavango Integrated Water Development Project,* IUCN, Gland.

Sears, J. (1988) Regional and seasonal variation in lead poisoning in the mute swan, *Cygnus olor*, in relation to the distribution of lead and lead weights in the Thames area, England. *Biological Conservation,* **46**, 115–134.

Segerstrom, U., Bradshaw, R., Hornberg, G. and Bohlin, E. (1994) Disturbance history of a swamp forest refuge in northern Sweden. *Biological Conservation,* **68**, 189–196.

Shaw, S. P. and Fredine, C. G. (1971) Wetlands of the United States, their extent and their value to waterfowl and other wildlife. *US Fish and Wildlife Service Circular,* **39**, 1–67.

Shay, J. M. (1984) Post fire performance of Phragmites australis. *Annual Report of the University of Manitoba Field Station (Delta Marsh),* **1983**, 18.

Skarén, U. (1988) Chlorinated hydrocarbons. PCBs and cesium isotopes in otters *Lutra lutra* from central Finland. *Annales zoologici fennici,* **25**, 271–276.

Skelton, P. H. (1990) The conservation and status of threatened fishes in southern Africa. *Journal of Fish Biology,* **37A**, 87–95.

Slater, F. M., Curry, P. and Chadwell, C. (1987) A practical approach to the evaluation of the conservation status of vegetation in river corridors in Wales. *Biological Conservation,* **40**, 53–68.

Smith, I. R. and Lyle, A. A. (1994) Running waters, in *The Fresh Waters of Scotland,* (eds P. S. Maitland, P. J. Boon and D. S. McLusky), John Wiley, Chichester, pp. 17–34.

Smith, P. G. R. and Theberge, J. B. (1986) Evaluating biotic diversity in environmentally significant areas in the Northwest Territories of Canada. *Biological Conservation,* **36**, 1–18.

Smith, B. D., Lyle, A. A. and Maitland, P. S. (1983) The ecology of running waters near Aberfeldy, Scotland, in relation to a proposed barytes mine: an impact assessment. *Environmental Pollution,* **32**, 269–306.

Smith, B. D., Maitland, P. S. and Pennock, S. M. (1987) A comparative study of water level regimes and littoral benthic communities in Scottish lochs. *Biological Conservation,* **39**, 291–316.

Smith, C., Youdan, T. and Redmond, C. (1995) Practical aspects of restoration of channel diversity in physically degraded streams, in *The Ecological Basis for River Management*, (eds D. M. Harper and A. J. D. Ferguson), John Wiley, Chichester, pp. 269–273.

Soutar, A. and Isaacs, J. D. (1974) Abundance of pelagic fish during the 19th and 20th centuries as recorded in anaerobic sediments off the Californias. *Fisheries Bulletin,* **72**, 257–273.

Stenmark, A. and Malmberg, O. (1987) (eds) *Parasites and Diseases in Natural Waters and Aquaculture in Nordic Countries*, University of Stockholm, Stockholm.

Stewart, B. A. and Davies, B. R. (1986) Effects of macrophyte harvesting on invertebrates associated with *Potamogeton pectinatus* L. in the Marina da Gama, Zandulei, Western Cape. *Transactions of the Royal Society of South Africa,* **46**, 35–49.

Steytler, N. S. and Samways, M. J. (1995) Biotope selection by adult male dragonflies (Odonata) at an artificial lake created for insect conservation in South Africa. *Biological Conservation,* **72**, 381–386.

Stoss, J. and Refstie, T. (1983) Short-term storage and cryopreservation of milt from Atlantic Salmon and Sea Trout. *Aquaculture,* **30**, 229–236.

Stott, B. (1977) On the question of the introduction of the grass carp (*Ctenopharyngodon* idella Val.) into the United Kingdom. *Fisheries Management,* **3**, 63–71.

Strahler, A. M. (1957) Quantitative analysis of watershed geomorphology. *Transactions of the American Geophysical Union,* **38**, 913–920.

Sukopp, H. (1971) Effects of man, especially recreational activities, on littoral macrophytes. *Hydrobiologia,* **12**, 331–340.

Swales, S. and Harris, J. H. (1995) The Expert Panel Assessment method (EPAM): a new tool for determining environmental flows in regulated rivers, in *The Ecological Basis for River Management,* (eds D. M. Harper and A. J. D. Ferguson), John Wiley, Chichester, pp. 125–134.

Tansley, A. G. (1939) *The British Islands and Their Vegetation*, Cambridge University Press, Cambridge.

Thieneman, A. (1912) Der Bergbach des Sauerlands. Faunistisch-biologische Untersuchungen. *Hydrobiologia (Supplement),* **4**, 1–125.

Thieneman, A. (1925) *Die Binnengewasser Mitteleuropas*. Schweizerbart'sche Verlagsbuchhandl, Stuttgart.

Thomas, G. L. (1982) Management of vegetation in wetlands, in *Managing Wetlands and Their Birds,* (ed. D. A. Scott), IWRB, Gloucester, pp. 21–37.

Thomas, J. W., Forsman, E. D., Lint, J. B. *et al. (1990) A Conservation Strategy for the Northern Spotted Owl,* United States Printing Office, Portland, ME.

Thompson, D. J. (1982) Effect of fire on *Phragmites australis* (Cav.) Trin. E. Stendel and associated species at Delta Marsh, Manitoba. University of Manitoba, MSc Thesis.

Tillyard, R. J. (1933) Mayflies of the Mt Kosciusko region, NSW: Introduction and Siphlonuridae. *Proceedings of the Linnaean Society of New South Wales,* **58**, 1–32.

Tilman, D. (1987) Secondary succession and the pattern of plant dominance along experimental nitrogen gradients. *Ecological Monographs,* **57**, 189–214.

Toha, J. C. and Jaques, I. (1989) Algal solution? *New Scientist,* **1691**, 71.

Tonn, W. M. (1990) Climate change and fish communities – a conceptual framework. *Transactions of the American Fisheries Society,* **119**, 337–352.

Tweed Foundation (1995) *Annual Report,* Tweed Foundation, Melrose.

Usher, M. B. (1986) Wildlife conservation evaluation: attributed criteria and values, in *Wildlife Conservation Evaluation,* (ed. M. B. Usher), Chapman & Hall, London, pp. 3–44.

Valverde, J. A. (1955) Essai sur l'Aigrette garzette (*Egretta garzetta)* en France. *Alauda,* **23**, 147–171; 254–279.

Verhoeven, J. T. A. and van der Toorn, J. (1990) Marsh eutrophication and wastewater treatment, in *Wetlands and Shallow Continental Water Bodies,* (ed. B. C. Patten), Academic Publishing, The Hague.

Wade, M. (1995) The management of riverine vegetation, in *The Ecological Basis for River Management,* (eds D. M. Harper and A. J. D. Ferguson), John Wiley, Chichester, pp. 307–313.

Walters, C. J. (1986) *Adaptive Management of Renewable Resources,* Macmillan, New York.

Ward, J. A. and Stanford, J. V. (eds.) (1979) *The Ecology of Regulated Streams,* Plenum Press, New York.

Welch, E. B. and Weiher, E. R. (1987) Improvement in Moses Lake quality from dilution and sewage diversion. *Lake Reservoir Management,* **3**, 58–65.

Whitman, W. R. (1982) Construction of impoundments and ponds at Tintamarre National Wildlife Area, Canada, in *Managing Wetlands and Their Birds,* (ed. D. A. Scott), IWRB, Gloucester, pp. 156–162.

Whitten, A. J. (1990) Recovery: a proposed programme for Britiain's protected species. *CSD Report No. 1089*, Nature Conservancy Council, Peterborough.

Wiese, J. H. (1976) Courtship and pair formation in the Great Egret. *Auk,* **93**, 709–724.

Williams, W. D. (1981) Running water ecology in Australia, in *Perspectives in Running Water Ecology,* (eds M. A. Locke and D. D. Williams), Plenum Press, New York.

Williams, P. H. and Gaston, K. J. (1994) Measuring more of biodiversity: can higher taxon richness predict wholesale species richness? *Biological Conservation,* **67**, 211–217.

Williams, J. E. and Miller, R. R. (1990) Conservation status of the North American fish fauna in fresh water. *Journal of Fish Biology,* **37A**, 79–85.

Wingfield, G. I. and Beeb, J. M. (1982) Simulation of deoxygenation of water containing vascular plants following treatment with terbutryne. *Proceedings of an International Symposium on Aquatic Weeds,* **6**, 255–262.

Wise, M. (1984) *The Common Fisheries Policy of the European Community,* Methuen, London.

Woodin, S. and Skiba, U. (1990) Liming fails the acid test. *New Scientist,* **10 March,** 50–54.

Wright, J., Moss, D., Armitage, P. D. and Furse, M. T. (1984) A preliminary classification of running water sites in Great Britain based on macroinvertebrate species and the prediction of community type using environmental data. *Freshwater Biology,* **14**, 221–256.

Wynne, G., Avery, M., Gubbay, S. *et al. (1995) Biodiversity Challenge: An Agenda for Conservation Action in the UK,* Royal Society for the Protection of Birds, Sandy.

Yalden, D. W. (1992) The influence of recreational disturbance on common sandpipers *Actitis hypoleucos* breeding in an upland reservoir in England. *Biological Conservation,* **61**, 41–49.

Yon, D. and Tendron, G. (1981) *Alluvial Forests of Europe,* Council of Europe, Strasbourg.

Zoltai, S. C., Pollett, F. C., Jeglum, J. K. and Adams, G. D. (1975) Developing a wetland classification for Canada, in *Forest Soils and Forest Land Management. Proceedings of the North American Forest Soils Conference, Quebec, 1973,* (ed. B. Bernier and C. H. Winget), vol. 4, pp. 497–511.

Index